IMAGES OF WAR SPECIAL

CENTURION
TANK

A Centurion armoured recovery vehicle (ARV, FV4006) photographed during the liberation of Kuwait in 1990/91. The registration number (00ZR48) indicates that this vehicle was converted from a Mk 1 or Mk 2 Centurion gun tank dating from the immediate post-war years. Note the additional composite armour applied to the sides of the vehicle in the form of panels. (*Tank Museum*)

IMAGES OF WAR SPECIAL

CENTURION TANK

RARE PHOTOGRAPHS FROM WARTIME ARCHIVES

Pat Ware

Illustrated by
Brian Delf

Pen & Sword
MILITARY

First published in Great Britain in 2012 by
PEN & SWORD MILITARY
an imprint of
Pen & Sword Books Ltd,
47 Church Street,
Barnsley,
South Yorkshire
S70 2AS

ISBN 978 1 78159 011 9

A CIP record for this book is available from the British Library.

Typeset by CHIC GRAPHICS

Printed and bound by CPI Group (UK) Ltd, Croydon, CR0 4YY

Pen & Sword Books Ltd incorporates the Imprints of
Pen & Sword Aviation, Pen & Sword Family History, Pen & Sword Maritime, Pen & Sword Military, Pen & Sword Discovery, Wharncliffe Local History, Wharncliffe True Crime, Wharncliffe Transport, Pen & Sword Select, Pen & Sword Military Classics, Leo Cooper, The Praetorian Press, Remember When, Seaforth Publishing and Frontline Publishing.

For a complete list of Pen & Sword titles please contact
Pen & Sword Books Limited
47 Church Street, Barnsley, South Yorkshire, S70 2AS, England
E-mail: enquiries@pen-and-sword.co.uk
Website: www.pen-and-sword.co.uk

Contents

Chapter One

Development

Although most people would rightly consider the A41 Centurion medium tank to be a post-war machine, the development process for the vehicle had actually started in the early autumn of 1943, more than twelve months before the D-Day landings. However, the first pre-production examples did not make it into Europe until after VE Day, and came too late to affect the outcome of the war. Nevertheless, it was clear from the outset that the Centurion was an excellent machine and, even with the original 17-pounder (76.2mm) gun, may well have been capable of standing up to the heavier German Tiger and Panther tanks. In fact, it would probably be fair to say that the A41 was the best tank of the Second World War that the British Army never had ... but it would be equally true to say that, despite being the final iteration of the flawed British cruiser tank concept, the Centurion went on to prove itself one of the best tanks of the immediate post-war period, regardless of origin. Notwithstanding its comparatively slow road speed and apparently insatiable thirst for fuel, it offered an excellent balance of firepower, protection and mobility, and was easily capable of being up-gunned.

The first Centurions entered service in December 1946, and the type saw its first combat in Korea in January 1951 with the 8th King's Royal Irish Hussars. By the time production came to an end in 1962, almost 4,500 Centurions had been constructed, and the vehicle saw service with nineteen armies across the world, notably also fighting in Aden, India/Pakistan, the Middle East and Vietnam.

To better understand the reasons behind the development of the Centurion it is necessary to look at the generally lacklustre performance of the tanks deployed by the British Army since 1939. These can be considered to fall into three categories: the pre-war tanks which the British Expeditionary Force (BEF) took to France in 1940, most of which were abandoned there after the evacuation from Dunkirk; the newer tanks developed for the British Army during the years 1940 to 1944; and the light and medium tanks supplied to Britain by the USA under the Lend-Lease arrangements. Despite a plethora of different machines, Allied tank design had generally failed to keep pace with developments in Germany since the mid-1930s, and almost all of the tanks fielded by the Western Allies during the Second World War were ill-matched to their German counterparts, in terms of both firepower and protection. The situation was not helped by the British Army's dogged insistence that tanks should be designed either as 'cruisers' or 'infantry tanks' – cruisers were described as fast, lightly armoured vehicles that could easily penetrate enemy lines, while infantry tanks were slower and more heavily armoured to suit the role of supporting advancing infantry. Even the iconic M4 Sherman medium tank, which was deployed in all of the major theatres of the conflict, was unable to meet the German heavy tanks such as the Tiger and Panther on equal terms.

By the time the Allies landed on the D-Day beaches in June 1944, the main British tanks in service were the A22 Churchill, which was officially described as 'infantry tank Mk IV', the A27M Cromwell, or 'cruiser tank Mk VIII', and the M4 Sherman, which, although not designed in Britain, was also considered to be a cruiser. None of the three was any match for the heavier German tanks and it was not until the appearance of the hybrid Anglo-American Sherman Firefly and the Cromwell-based Comet, both of which were armed with a British 17-pounder (76.2mm) gun, that the Allies were able to field a tank that was able to pose any real threat to the opposition. However, neither had been designed from first principles and both could be considered to be 'what-if' developments of earlier designs that were inevitably compromised in various ways.

With the war showing every sign of dragging on well into 1945, it had long been obvious that both Britain and the USA would require better-armed and more heavily armoured tanks to defeat the German Tiger and Panther. In the USA, work had been progressing on a replacement for the Sherman since 1942. This eventually appeared as the M26 Pershing at the beginning of 1945. In Britain, meanwhile, work on the Centurion – originally simply designated A41, the name Centurion having previously been assigned to the A30 tank that actually ended up being called Challenger – had started in the summer of 1943, initially with a view to seeing full-scale production under way by November 1945.

By October 1943 the British Tank Board had met to consider a paper that had been drawn up by the Army Council Secretariat describing the desirable characteristics of future cruiser tanks. Embodying all of the lessons in tank design that the British Army had learned fighting the Germans in the Western Desert, the paper was intended to provide a starting point for the development of a new heavy cruiser tank. Recognising that the existing process of allowing industry to design new tanks had not proved particularly successful, and in the light of criticisms in Parliament and in the press with regard to the weaknesses of existing tanks in terms of firepower and reliability, it was decided that the Ministry of Supply's Department of Tank Design (DTD), led by A.A.M. Durrant, would be tasked with preparing the outline design for what was being described as the 'A41 heavy cruiser tank'. In time, the DTD would become the Fighting Vehicles Design Department (FVDD), before merging with the

Fighting Vehicles Proving Establishment (FVPE) to become the Fighting Vehicle Research & Development Establishment (FVRDE).

Major emphasis was to be placed on firepower and protection, even if this was at the expense of mobility, and the new tank was intended to match the armour, firepower and automotive performance of the 75mm-equipped German Pz Kpfw V Panther in all practical respects.

A document was prepared that spelt out the major design features. For example, it was stated that the turret ring should not be less than the 69in diameter of the American Sherman. The new tank was to be armed with the 17-pounder (76.2mm) quick-firing anti-tank gun that had proved so effective against German armour in the Sherman Firefly ... but there was some suggestion that this might eventually be replaced by a huge 37-pounder gun firing separated ammunition, a proposal which came to naught! The armour was to be sufficient to withstand the powerful German 88mm KwK 36 and PaK 43 guns, and the hull design was to incorporate a sloping glacis plate to improve frontal protection. For the first time in a British cruiser tank, the hull machine gun was said to be unnecessary, which allowed the number of crew to be reduced to four. The overall weight was to be constrained to a maximum of 40 tons – although this was subsequently to prove impractical and the weight limit was increased to 47 tons – and the overall width was not to exceed 126in, the latter being intended to allow the tank to be transported by rail, and to cross a standard Bailey bridge. A high road speed was not considered to be important since combat experience had generally shown that it offered little advantage, although cross-country performance was to be at least as good as that of the Comet and Cromwell tanks, and a high-speed reverse gear was considered to be essential. An initial life-mileage of 3,000 miles was demanded, along with an optimum balance of reliability, and simplicity of operation and maintenance.

Power was to be provided by the Rolls-Royce V12 Meteor petrol engine in Mk 4 guise. Based on the iconic Merlin aircraft engine, but lacking the supercharger that had boosted the Merlin's power output to more than 1,000bhp, the Meteor had been developed by W.A. Robotham at Rolls-Royce in 1941 expressly for use in tanks. It was originally used to replace the ageing Liberty V12 engine that dated back to 1916, and had already been installed in the Challenger, Comet and Cromwell to huge success, proving itself to be both powerful and reliable.

At this stage in the war, and particularly bearing in mind the criticisms that had been levelled at the propensity of the Sherman to catch fire when hit, the choice of a petrol engine might seem strange. However, the decision was at least partly determined by the logistic use of fuels in the British services: the Royal Navy had priority on diesel fuel and the RAF required high-octane aviation spirit, leaving the army reliant on petrol ... despite the very real risk of fire when used in armoured fighting vehicles. It was felt that this arrangement allowed the largest possible quantity of fuel to be obtained from a given quantity of crude oil and, although these days the stated policy is now to use diesel fuel in all armoured vehicles, the decision to continue to use petrol in fighting vehicles was upheld in 1951 and remained policy for a further decade or more.

Unlike the Merlin, where weight was a critical issue, the Meteor used cast-iron wherever possible and, in order that it could be a direct replacement for the Liberty engines that had been fitted to British tanks such as the Crusader, Centaur and Cavalier, the overall height of the engine was reduced by some six inches when compared to the Merlin, by repositioning the water and oil pumps, and various accessories. The direction of rotation was also reversed, with the Meteor running clockwise when viewed from the nose; other changes from the original Merlin engine included the use of detachable cylinder heads. However, a large percentage of the parts were interchangeable between the two engines, including the crankshaft and connecting rods.

The engineers at Rolls-Royce's Clan Foundry in Belper, Derbyshire, had access to stocks of early Merlin engines and some of these were used to produce the first mock-ups of the Meteor. However, the adoption of the Merritt-Brown gearbox forced a number of design changes which meant that many of the Merlin castings were no longer suitable, and the first 'proper' Meteor engine was installed into a Crusader tank (number T15646) in April 1941. By July Rolls-Royce had completed three more engines and had shipped them to the company's Pym's Lane factory at Crewe for testing. Once the design had been proven, it was planned that Leyland Motors would start to construct 1,200 engines for the Cavalier Mk VII tank, with production to reach 500 a month by autumn 1942. However, nervous that the more powerful Meteor engine would highlight weaknesses in other areas of the Cavalier, particularly the ability to cool the engine effectively, and having tooled up to continue producing Liberty engines, Leyland backed out, leaving Rolls-Royce to produce the first 1,000 engines, with the Ministry of Supply subsequently increasing the requirement to 3,000.

It was not an ideal arrangement. Rolls-Royce, already busy producing the Merlin, lacked sufficient capacity to also produce the numbers of Meteors that would eventually be required. It was subsequently agreed that Rolls-Royce would maintain its design parentage, and would produce the components for the engines, with the Wolverhampton-based Henry Meadows company undertaking assembly, occasionally using parts salvaged from the Merlin engines of crashed aircraft! Meteor engines were also subsequently produced by the Nuffield Mechanizations division of Morris Motors, but by 1943 production and design parentage of the Meteor had been passed to the Rover Company at Solihull in 'exchange' for the gas turbine engine designs that Rover had been working on for the Ministry of Supply. By the time the war was over, production of the Meteor, and the subsequent V8 Meteorite, resided only with Rover, which continued to manufacture the engine at Drakelow and then at Tysley until 30 June 1964, when the very last one was built. For the Centurion development project, Rover was initially commissioned to provide eighteen Meteor Mk 4 engines with a power output of 550bhp, but by the time the development process was finished this figure had risen to 650bhp.

In what was something of an innovation for British tanks, the Centurion was also to be fitted with an auxiliary engine coupled to a generator, allowing the radios, turret traverse and gun-control equipment, as well as the all-important tea-brewing equipment – coyly described as a 'boiling vessel' – to continue to operate without any need for the main engine to be running.

The transmission was a redesigned version of the Merritt-Brown five-speed combined gearbox and steering unit that had already been well proven in the Cromwell and Comet tanks. The major differences when compared to earlier versions of this transmission were the inclusion of a high-speed reverse gear, capable of propelling the Centurion backwards at 7.6mph, and an automatic differential lock. The track width was initially specified at 24in.

Finally, the increased weight of the vehicle when compared to earlier cruiser tanks meant that the old Christie suspension had to be abandoned. AEC's designer, G.J. Rackham, came up with a modified version of the Horstman system that it was felt would provide the best possible cross-country performance. Although Rackham introduced a number of new features, notably the use of three concentric coil springs, the Horstman suspension system was not new. It was originally designed by Sydney Horstman, the son of a German watchmaker who had settled in Bath in 1854, changing his name from Horstmann after the end of the First World War, and had first been used on the Vickers Mk II light tank back in the mid-1930s. In redesigning Horstman's system, Rackham made certain that there was sufficient allowance to accommodate subsequent increases in weight as the tank was inevitably up-gunned and upgraded during its life.

A design team for the tank was established under the chairmanship of Sir Claude Gibb, the Director General of Tank Production. An outline specification and general arrangement drawings had been completed by October or November of 1943, with a full-size mock-up available by the end of the year. Following a protracted discussion, it was decided that pilot and pre-production models would be constructed towards the end of 1944, and AEC, which had been building engines for the Valentine tank, was appointed as the 'design parent'.

By late May 1944, scarcely two weeks before D-Day, the AEC workshops in west London, working in conjunction with the Royal Ordnance Factory (ROF) at the

(*Above*) The original caption to this photograph described it as the 'A41 prototype'. This view across the rear deck shows the distinctive cooling louvres, the exhaust silencers mounted on the track guards and the circular escape hatch at the rear of the welded turret that was also designed to allow the barrel of the 17-pounder (76.2mm) gun to be removed. The stencilled legend on the side skirt indicates that anti-freeze has been added to the cooling system. (*Warehouse Collection*)

(*Opposite, top*) The same vehicle from the front, showing the 17-pounder (76.2mm) gun and the 20mm coaxial Polsten cannon. The warning triangle, just visible on the turret side, normally indicates that the component is not made from armoured steel. Records show that the vehicle was delivered in April 1945. (*Warehouse Collection*)

(*Opposite, below*) The legend 'P1' on the turret side indicates that this is 'prototype number one' again, but the turret has now had the distinctive stowage bins attached to it, and spare track links are carried on the upper glacis plate. The upstanding brackets that were used to attach the side skirts were unique to the prototypes, as were the narrow 20in-wide tracks. (*Warehouse Collection*)

Woolwich Arsenal way over on the south-east side of the capital, had produced the first mock-up of the hull in mild steel. Although lacking both gun and turret, the intention was simply to 'prove' the choice of suspension, engine and transmission, and the control systems. A second hull, this time of armoured steel, was also produced for stowage and defensive firing trials. It was intended that these mock-ups would be followed by twenty prototypes, or pilot models, to be constructed at ROF Woolwich and Nottingham, with the Centurion entering full production by November 1945.

At this stage, the definitive arrangement of the Centurion's armaments had yet to be agreed and four different specifications were proposed. Prototypes numbered one to five were to be armed with a 17-pounder (76.2mm) L/55 gun, together with a coaxial 20mm Polsten machine gun in the turret, a Polish development of the Swiss Oerlikon 20mm automatic cannon, that was felt to be capable of destroying unarmoured targets, together with a rear-facing 7.92mm Besa machine gun, the latter being a British version of the Czechoslovak ZB-53 air-cooled belt-fed machine gun. Prototypes numbered six to ten also had the 17-pounder (76.2mm) main gun and the 20mm Polsten machine gun, but the rear-facing Besa gun was omitted in favour of a circular escape hatch that also doubled as a means of extracting the gun barrel. On prototypes eleven to fifteen the Polsten was replaced by a coaxial 7.92mm Besa machine gun. And, finally, prototypes numbered sixteen to twenty were to be fitted with the dual-purpose L/49 77mm main gun that had been fitted to the A34 Comet tank (and was effectively a shortened version of the 17-pounder), together with a 7.92mm Besa machine gun.

Some period documentation also suggests that other differences between these initial twenty tanks and those that followed included a track width of 20in, rather than the standard 24in, and the glacis plate thickness being 57mm thick rather than 76mm.

Although it was almost certain that the Merritt-Brown controlled-differential steering and transmission system would be specified for production, it was also decided that the last five prototypes would be fitted with a Sinclair-Meadows Powerflow SSS ('synchromesh self-shifting') twin-range automatic transmission, produced by the Hydraulic Coupling & Engineering Company. Offering four forward gears and three reverse, with a maximum speed in reverse of 14mph, the Powerflow transmission operated via a fluid clutch, and was fitted with a so-called auxiliary 'traction gear' that maintained the drive to the tracks during the gear-changing process. This minimised any loss of forward (or, for that matter, reverse) momentum. It was an unusual transmission system that at one time had been considered as a replacement for the Wilson pre-selector gearbox that had been fitted to various British tanks both before and during the Second World War.

The Centurions that were scheduled to be fitted with the Powerflow system were to be designated A41S, but in the event only one vehicle (prototype number five) was actually equipped with this transmission and it quickly became clear during trials that it was not sufficiently reliable. One gearbox was destroyed after little more than 500 miles, causing considerable collateral damage, and although its replacement completed more than 1,000 miles, it suffered frequent failures during the trials, requiring many components to be replaced. Even when operating correctly, the Powerflow transmission was not popular with drivers since it required an unorthodox driving technique, and eventually it was discarded as unsuitable.

The first pilot Centurion was completed by ROF Woolwich in April 1945 and was delivered to the Fighting Vehicles Proving Establishment (FVPE) at Chertsey, Surrey, for automotive trials, under the direction of Brigadier W. Morrogh. Two more followed, and the vehicles acquitted themselves extremely well during the trials, the first covering 1,055 miles, of which almost half consisted of off-road running. The maximum road speed attained during the trials was 23.7mph. There is some doubt as to whether all of the planned twenty prototypes were actually constructed and the number may actually have been just seventeen, but a fourth vehicle was certainly delivered to Lulworth for gunnery trials during April and by the following month, May 1945, the total number of vehicles available was six. Of these, three had been completed by ROF Woolwich (prototypes three, nine and eleven, numbered T352412, 352416 and T352417, respectively), and three by ROF Nottingham (prototypes four, six and eight, numbered T352413 to T352415).

During April 1945, as part of a broader trials programme, it had been decided that the six tanks would be shipped to north-west Europe in order that they could be tested under active service conditions. The original idea had been to crew the tanks with men drawn from the Grenadier, Coldstream, Welsh and Irish Guards, under the command of Captain Sir Martin Beckett MC of the Welsh Guards. The hope was that the Centurions could be actually used in combat against the Wehrmacht, but all of this changed after Germany's surrender in May. Although the trials still went ahead during June and July 1945, they were of a somewhat different nature and the men that were used to crew the tanks formed a Guards detachment attached to 7th Armoured Division.

Six Centurions were shipped from Southampton by landing craft on 14 May 1945, which presumably also resulted in some other useful operational information regarding the deployment of the new tank by landing craft. Although there is no suggestion that the tanks did anything other than sail directly to Antwerp, nevertheless they did not arrive until some five days later. Once on dry land, the Centurions undertook a 400-mile 'march' across Belgium and the Netherlands to Gribbohm, in the German province of Schleswig-Holstein. During the troop trials, which were conducted under the codename Operation Sentry, particular emphasis was placed on comparing the Centurion to the Sherman and the Cromwell under both road and cross-country conditions. However, there was no opportunity to actually compare the Centurion to the German tanks in real-life combat situations. The troop trials were followed by gunnery trials at the Lommel range in Belgium's north-eastern Limburg province, close to the border with Germany and the Netherlands. With the trials over, the tanks were shown to other armoured regiments in Europe before being returned to Britain via Calais in July.

Despite the failure of one engine and three gearboxes, and a propensity for the Centurion to shear the front idler brackets, the tanks proved reliable and were popular with the crews. Many suggested that the Centurion was the best tank they had seen, although it wasn't successful in every respect; during swamp-crossing trials, for example, carried out under the eagle eye of the 'Mud Committee', the Centurion was shown to be inferior to the Cromwell and the American M24 Chaffee. Other criticisms included the Polsten gun, the weapon being singled out for adverse comment because it apparently protruded too far into the turret and was not always reliable. Weight of opinion suggested that the Polsten should be replaced by a 0.30in Browning which would release some space, as well as having the advantage of using standard NATO ammunition. Likewise, it was also suggested that the Besa machine gun that was fitted to prototype number eleven should also be replaced by a Browning ... and there was also some pressure to fit a Browning on the commander's cupola.

During August 1945 22nd Armoured Brigade submitted a list of minor recommendations for changes and modifications, but there was nothing of any real substance and, following discussions at the 22nd meeting of the DRAC's (Director, Royal Armoured Corps) Advisory Committee, it was agreed that contracts would be issued for an initial quantity of 800 Centurions. One hundred of these were to be of the Centurion Mk 1 (A41*) configuration, armed with the 17-pounder (76.2mm) L/55 main gun together with a linked, but not coaxial, 7.92mm Besa machine gun. (See colour plate 1, a and b.) Construction of the Mk 1 was allocated to ROF Woolwich, Nottingham and Leeds (Barnbow), although apparently the latter, which had only been converted for tank production at the end of the war, was to contribute just three vehicles.

Of the remaining tanks, the next hundred were described as Centurion Mk 2 (A41A), and were to be equipped with the 17-pounder gun in a cast, rather than welded rolled-steel turret incorporating a commander's cupola, a combined gunner's periscopic sight, and a coaxial machine gun. Although new, the design of this turret had been in hand for some time. The remaining 600 tanks, which became known as the Centurion Mk 3, were to be equipped with the new 20-pounder (83.4mm) L/66.7 gun that was still under development. (See colour plate 2, a and b.)

Centurions Mk 2 and 3, which would be constructed by Vickers-Armstrongs at Elswick and ROF Leeds, were also to incorporate improvements to the engine, which was now described as the Meteor Mk 4A. The power output was up from 550bhp to 600bhp by virtue of increasing the compression ratio from 6:1 to 7:1. At the same time, the final-drive ratio was altered to compensate for the increase in weight, bringing the maximum theoretical road speed down to 21.4mph from the previous 23.7mph.

Meanwhile, although no production vehicles were yet available, acceptance trials for what would be the Centurion Mk 1 had been put in hand during 1946 using prototype number twelve. This had been reworked to what was described as 'Centurion I specification' by 'the erector' (either ROF Woolwich or ROF Nottingham). However, this reworking did not extend to replacing the Metropolitan-Vickers traversing gear with the Lucas equipment that had been selected for the first production models. The acceptance trials were planned to cover 3,000 miles, and the report produced at the end of the trials period described reliability over the first 1,000 miles as 'satisfactory', with just three major defects coming to light, none of which would have prevented the tank from remaining in operation. There was also a list of some forty-seven minor defects, a number that was considered to be more than would normally have been experienced, but some of the problems arising during the trial were already known about, and steps had already been taken to correct them on production vehicles.

One design point that was singled out for attention related to the cooling system. Because the main and auxiliary engines shared the cooling circuit, it was important that the main engine was started first, and failure to observe this simple precaution could, under some circumstances, result in overheating of the auxiliary engine with resulting damage.

During the trials the average road speed of the tank, measured over 100 miles, was 19.9mph, or 86 per cent of the governed speed; across country the average speed was reduced to 12.6mph, largely in deference to the comfort of the crew, and factors such as the ride were noted as 'good', with the suspension described as soft and well damped. Fuel consumption was recorded as 0.83mpg on the road, reducing to 0.38mpg across country, and oil consumption was measured at a staggering 39mpg ... and if that figure is correct it would almost suggest that the Meteor was a two-stroke! Although these figures were considered to be 'satisfactory', comparing favourably with other prototype vehicles tested under similar conditions, the heavy fuel consumption remained one of the Centurion's weak spots throughout its life. There were no firing trials at this stage.

Many were of the opinion that the Centurion deserved the title of 'best British tank', if not, at that precise moment, the 'best tank in the world'. However, despite all

The Centurion Mk 2 (A41A) was equipped with the 17-pounder (76.2mm) gun in a cast, rather than welded, rolled-steel turret that incorporated a commander's cupola, a combined gunner's periscopic sight and a coaxial machine gun. (*Warehouse Collection*)

this praise and notwithstanding that the Centurion had acquitted itself well in almost every practical respect, it appeared that the service life of the vehicle might be short-lived. The reason for this was contained in a document that was drawn up in 1947 and proposed that, for the future, the British Army be equipped with three types of tank: the 100-ton FV100 assault tank, the 50-ton FV200 'universal' tank and the 10-ton FV300 light tank. This did not include the Centurion and, while the FV100 and FV200 were to be based on the Centurion running gear, the Centurion itself was now being viewed as a stopgap measure that would be superseded as the new tanks entered service. As is so often the case, this programme was wildly optimistic. The FV100 and the FV300 were soon abandoned and, although the FV200 'universal' tank – later to become known as the Conqueror – did finally enter production, fewer than 200 were constructed and it was forced to serve alongside the far more adaptable Centurion, which had received official sanction in 1950. It could be argued that the Conqueror, which was a response to the massive Soviet IS-3 tank that had appeared in 1945 and was armed with a 122mm gun, was a victim of its own considerable size and weight. It was never felt to be satisfactory and by the late 1960s both the Conqueror and the Centurion had been replaced by the Chieftain ... which followed many of the successful design principles of the Centurion.

So the Centurion remained in production and by 1948 ROF Leeds, ROF Dalmuir and Vickers-Armstrongs had started production of the Centurion Mk 3. Armed with a new 20-pounder (83.4mm) L/66.7 gun, designated 'tank gun, Mk 1', in place of the old 17-pounder, the tank embodied dozens of additional modifications. Many of these had arisen out of experience gained while operating the Mk 1 and Mk 2 variants, and the modifications included the use of more advanced Metropolitan-Vickers electric and gyroscopic gun-control equipment that allowed the tank to fire on the move, with the Centurion being the first British tank to be so equipped. There was also revised ammunition stowage, with sixty-five rounds carried rather than seventy-four since the 20-pounder rounds were larger than those of the 17-pounder. The Mk 3 was powered by the Mk 4B version of the Meteor engine, with the power output now increased to 620–630bhp at 2,400rpm.

The Centurion Mk 3 was the first of the type to see active service, fighting in Korea from 1950 in support of United Nations troops. In combat, it quickly proved itself superior to other tanks in the theatre, including the old Soviet T-34/85s dating from the Second World War. Although it was generally well received, by 1952 a head of criticism had built up regarding the thickness of the Centurion's floor, which was said to be easily distorted by the anti-personnel mines that were being laid by the Chinese. It was suggested that a second, false floor be fitted to help deflect the blast, but since the first such mine that the tank encountered would cause the false floor to collapse there would be no further beneficial effect. Needless to say, the idea was

Still using the pre-1949 style of census number, T351970 is a Centurion Mk 3 with the cast turret and 20-pounder (83.4mm) gun; the gun mantlet is now fitted with a protective canvas cover, through which the coaxial machine gun projects. The attachment brackets and bolts for the side skirts – or 'bazooka plates' – no longer project above the track guards and the tank is now running on the standard 24in-wide tracks. (*Warehouse Collection*)

Rear view of Mk 3 T351970 showing the spring-mounted towing pintle, 24in wide tracks and the prominent recovery cable. (*Warehouse Collection*)

dropped and the 17mm thickness of the floor remained standard throughout the life of the Centurion. However, by 1950 some 250 individual improvements had been made to the basic design, including the relocation of the loader's hatch to provide easier access and escape, and the deletion of the rear escape hatch in the turret that had been intended to allow the 17-pounder gun barrel to be removed through the interior of the tank. Most Mk 3 Centurions were subsequently upgraded to Mk 5 standard.

If the Centurion did suffer from one particular defect, it was that of thirst. The power to weight ratio was low, and the Meteor engine consumed fuel at an eye-watering rate – just 0.27–0.52mpg – giving a range of just 33 to 62 miles from the 121-gallon tanks. During the Soviet blockade of the city of Berlin in 1948 it became clear that the Centurions would not be able to reach the city from their bases in West Germany without having to refuel en route, which was hardly an ideal situation!

While improvised external auxiliary fuel tanks were frequently fitted to improve the range, they were, on occasion, more of a hindrance than a help. For example, a pair of externally fitted 44-gallon drums introduced a hazardous and vulnerable factor in combat. Being essentially unarmoured, they were very susceptible to damage, and could often result in fire. In 1952 the Chertsey-based Fighting Vehicles Research & Development Establishment (FVRDE) came up with a mono-wheeled 200-gallon armoured fuel trailer (FV3751), designed to be hitched to the rear of the tank by a pair of side arms, and able to be jettisoned by means of explosive bolts when it was no longer required. Weighing around 2 tons when full of fuel, the trailer ran on a single trailing wheel shod with a run-flat tyre and carried on a pair of coil-sprung wishbones. Fuel was transferred from the trailer to the twin fuel tanks of the towing vehicle by two electric self-priming pumps, pushing the fuel through rubber hoses that were fitted with self-sealing couplings. Manufactured by Joseph Sankey, and eventually used by the British Army and by the armies of Sweden and the Netherlands, the trailer improved the operating range by around 50 to 100 miles, but by all accounts it was singularly unpopular with the Centurion crews by virtue of the manifold difficulties involved both in attaching the twin side arms of the trailer to the tank, and in manoeuvring the tank with the trailer attached – it was apparently all too easy to reverse over the trailer during a ditch crossing, or in very rough country, with disastrous results. Several Centurions were also burned-out as a result of fires caused by fuel being pumped from the trailer while the main fuel tanks were full, causing the excess fuel to spill out of the breathers into the engine compartment, where it was ignited. The use of the trailer was finally abandoned in 1963 when an additional 100-gallon armoured fuel tank was bolted to the rear of the hull.

Logic would dictate that the next variant to appear would be the Centurion Mk 4. It had always been intended that there would be a close-support variant of the tank, equipped with a 95mm 'tank howitzer, Mk I' capable of firing high-explosive and smoke rounds, and it was this version that was designated Mk 4. So important was this role considered that at one time it was being suggested that 10 per cent of all Centurions constructed would be armed with the 95mm gun. Following the construction of a wooden mock-up for appraisal, in 1947 a Mk 1 was rebuilt to meet the Mk 4 specification. It was planned that ROF Leeds would convert Centurion Mk 1s to Mk 4 configuration and some work even started on the production of some of the components required, but in the event it was felt that the 20-pounder (83.4mm) gun was sufficiently powerful to fulfil the close-support role and the War Office cancelled the project in 1949 with no production vehicles constructed. The Mk 4 designation was not re-used to describe any other variant.

In late 1952, in response to experiences gained in Korea, development work started on the Mk 5 variant (designated FV4001), now under the design parentage of Vickers-Armstrongs rather than the Department of Tank Design, with production eventually getting under way in 1955/56. (See *colour plate 3, a and b*.) Compared to the earlier Mk 3, the changes were minimal, but the coaxial Besa machine gun was replaced by an 0.30in Browning, the loader's 2in bomb thrower was eliminated, and his roof-mounted periscope was relocated; a cupola-mounted machine gun was subsequently included and changes were made to the sighting and fire-control equipment. An additional single-width track guide roller was also fitted between the idler wheel and the first return roller, and between the drive sprocket and the last return roller. These changes were also made to the Mk 3, bringing most existing vehicles up to Mk 5 specification. Other amendments apparent on the Mk 5 included reshaping of the turret roof and the omission of the rear escape hatch in the turret. Late production Mk 5 variants were fitted with what was described as the 'B-barrel', which was fitted with a fume extractor – sometimes described as a 'bore evacuator' – mounted mid-way along the length of the barrel. The purpose of the fume extractor was to prevent the hot gases and other combustion products created when the gun was fired from leaking into the fighting compartment, by taking combustion gases into the extractor as the shell passed through it, and subsequently releasing them back into the barrel once the projectile had exited.

The Mk 5/1 (FV4011) had additional armour welded to the glacis plate, and the Mk 5/2 was the first variant to be fitted with the 105mm L7 rifled gun – the 'L' aspect of this nomenclature should not be confused with the 'L' shorthand used to denote barrel length as part of the barrel length-to-calibre ratio. Work on the new gun had started in 1956, and early versions compared well in firing trials against the American 120mm gun, and showed a near 25 per cent improvement in penetrating power over the old 20-pounder. The first user trials of the new gun were carried out during July 1959 on the British Army ranges at Hohne in Germany, using a pair of Mk 8

A late model Centurion Mk 3 being loaded onto the FV3001 Sankey 60-ton tank-transporter semi-trailer. The potential difficulties associated with the mono-wheeled fuel trailer are more than apparent in this view, which shows it suspended in midair. (*Warehouse Collection*)

An adaptor was developed for the FV3751 mono-wheeled fuel trailer that allowed it to be towed by a standard 1-ton truck. (*Warehouse Collection*)

Centurions that had been fitted with the 105mm gun by the Royal Electrical & Mechanical Engineers (REME). The 105mm L7 went on to be adopted as the standard NATO tank gun. As well as being produced in Britain, it was also manufactured in the USA as the Watervliet Arsenal M68, and in West Germany as the Rheinmetall Rh-105-60; it also prompted development of the 105mm GIAT CN105 gun in France and the OTO Melara 105mm gun in Italy. The Centurion was the first practical application of the weapon, but it was also used to equip the German Leopard 1, the Israeli Merkava, the US Army's M60 Patton and M1 Abrams, and the Vickers MBT and Valiant tanks. It was a fine weapon, hard hitting, reliable and accurate ... in fact, it was so accurate that by 1966 British Centurion gun crews had twice won a competition, initiated in 1963, to test the speed and accuracy of tank gunnery.

The Mk 6 was a conversion of existing Mk 5 tanks, up-armoured and fitted with the 105mm L7 gun. From about 1962 a 22in diameter dual-purpose white light/infrared searchlight, with a range of about 500 yards, was mounted on the gun mantlet, and aligned with the gun barrel to provide a night-fighting capability. The commander was provided with an appropriate infrared periscope that could be fitted in place of the standard item when required. At the same time, infrared driving lights were fitted alongside the standard headlamps. This modification was indicated by adding the suffix 'I' to the basic 'mark' number, to give Mk 6/1. The Mk 6/1 was also provided with an open stowage basket on the turret rear to carry a camouflage net, thus avoiding the need to carry it over the engine grilles, a practice which had been found to lead to overheating. The Mk 6/2 was the first Centurion variant to be fitted with an 0.50in coaxial range-finding machine gun mounted in the gun mantlet, as well as having increased fuel capacity, the latter occasionally being indicated by the suffix 'LR' (Centurion Mk 6/2 LR). At the same time, a thermal sleeve was added to the barrel of the 105mm L7 gun to reduce distortion due to heating during sustained firing.

By the time the much-improved Mk 7 (FV4007) was introduced in 1953/54, design parentage had moved again, this time to Leyland Motors, with production taking place at a brand-new Ministry of Supply factory at Preston, as well as at ROF Leeds (Barnbow). (*See colour plate 6, a and b.*) The design specification for the Mk 7 was published in October 1952, with production scheduled to begin later that year; all of the previous modifications were incorporated, the internal arrangements were improved, and the turret turntable was fitted with a rotating floor. This was also the first variant to be fitted with headlamps. At first, the 20-pounder gun was retained, but changes included increased stowage for ammunition, a contra-rotating commander's all-round vision (ARV) cupola, and a fume extractor on the 20-pounder barrel. The hull was redesigned at the rear to include provision for an additional armoured fuel tank that provided a near-doubling of the fuel capacity without resorting to the use of trailers or auxiliary fuel tanks. At the same time, the threaded fastenings used on the tank were changed from the old BS 'fine' (BSF) and BS 'coarse' standards (the latter described as BSW or 'Whitworth') to the so-called 'unified' threads (UNF and UNC) that had been adopted for all NATO equipment to improve interoperability.

Automotive trials were conducted under the auspices of FVRDE at Chertsey, where the tank was run with both Mk 3 and Mk 7 turrets. Despite a gearbox failure at 2,001 miles, and problems with the cooling system, the latter apparently arising as a result of the changes made to the fuel filler caps, these trials showed that both performance and reliability were considered to be generally satisfactory. Further trials were conducted at the Royal Armoured Corps (RAC) Equipment Wing, using Centurions 42BA28 and 42BA39, with a final report published in November 1955. The RAC trials had started in August 1954, with 42BA28 initially being sent for gunnery trials, and 42BA39 being retained for automotive trials. On completion of the gunnery assessment and firing trials, where the tank was initially described as 'operationally unfit due to defects in the breech and failure of the coaxial machine gun' due to it jamming every twenty rounds, 42BA28 was returned to Bovington where it was put through automotive trials covering 2,000 miles. Meanwhile, 42BA39

This late model privately owned Centurion (Mk 6/2 or Mk 9/2) is equipped with the 105mm L7 gun, to which has been fitted a thermal sleeve, and ranging machine gun. (*Simon Thomson*)

From about 1962 Mk 6 Centurions (and later variants) were fitted with a 22in diameter dual-purpose white light/infrared searchlight, with a range of about 500yd. The searchlight was mounted on the gun mantlet and aligned with the gun barrel to provide a night-fighting capability. (*Warehouse Collection*)

completed an initial 2,011 miles before being stripped down to assess levels of wear; after reassembly, the trials continued to a total of 3,168 miles. Eventually, both vehicles were described as 'acceptable' subject to the commander's binoculars being graduated in 'mils' (one mil being equal to 1/6,400 of a complete revolution). It was also noted that 'weaknesses in the unified threads' needed to be rectified. During the trials, this weakness had led to the brake drums working loose and the armoured side skirts actually falling off. It was also suggested that steps should be taken to prevent the possibility of the magnetos accidentally being switched off, and that a hatch should be provided to aid topping-up the engine oil.

Following the completion of the initial phase of these trials, it was proposed that the Meteor Mk 4C engine be adopted, with shaft-driven cooling fans and a new generator mounting ... and that the trials be continued for a further 2,000 miles.

On the downside, it was reported that the increase in the vehicle weight resulting from the additional fuel being carried – the weight of the fuel alone was somewhere in the order of 850lb – had the effect of reducing the life of the tyres on the rear-most road wheels by some 40 per cent. However, this was felt to be a price worth paying for the increase in range, which was almost doubled. A pair of Mk 7s were also tested at the British Army Training Unit, Suffield (BATUS) in Canada and, as a result, the track shoes were redesigned to include a new transverse bar that improved grip in ice and snow.

The Mk 7/1 (FV4012) was an up-armoured Mk 7, and the Mk 7/2 was fitted with the 105mm L7 gun.

Centurion Mks 8 (FV4014) and 8/1 (FV4007) went into production in 1955/56, retaining the 20-pounder gun but fitted with a new gun mantlet having resiliently mounted trunnions that reduced the possibility of serious damage to the gun mount or turret in the event of a direct hit on the mantlet; the new mounting omitted the canvas cover and necessitated some redesigning of the turret castings. The turret roof incorporated a new contra-rotating commander's cupola with twin lightweight hatches. There was also a new gun-elevation system using chain drive rather than gears, and improved gun-control equipment including an emergency gun-firing system. The engine and transmission covers were also redesigned to improve cooling performance.

As a result of the appearance of the Soviet T-54 tank in Hungary in 1956, at least one example of the Mk 8 was fitted with an additional 51mm of appliqué armour on the glacis plate, giving a total of 127mm, compared to 120mm on the T-54. This modification was adopted for production, leading to the designation Mk 8/1, and the threat posed by the T-54 was such that additional armour was retrospectively fitted to Mk 5 and Mk 7 tanks at their next base overhaul at a cost of £35,200 per tank.

The Mk 8/2 was fitted with the 105mm L7 gun, as were all subsequent marks.

Just one example of the Mk 9 (FV4015) was constructed during 1959/60. It was effectively an up-armoured Mk 7 fitted with the 105mm L7 gun, but there was no series production. However, 200 conversion kits were subsequently manufactured which would allow Mk 7s to be converted to Mk 9 (and Mk 9/1 and Mk 9/2) configuration, a task requiring 130 man-hours of work. The Mk 9/1 (FV4007) also had infrared fighting and driving equipment and was fitted with the turret stowage basket, while the Mk 9/2 (FV4007) had the 0.50in ranging machine gun.

Also dating from 1959/60 was the Mk 10 (FV4017), the final production variant. It was effectively an up-armoured Mk 8 fitted with the 105mm L7 gun at the factory, but there was also increased stowage for ammunition for the main gun and a new design of bearing in the final-drive assembly better able to resist shock loads. The main identifying feature was the revised louvres fitted over the engine deck, which were now of a chevron pattern. The Mk 10/1 was fitted with infrared fighting

Centurion Mk 10 with the turret traversed to the rear. The Mk 10 (FV4017) was effectively an up-armoured Mk 8, fitted with the 105mm L7 gun; note the fume extractor on the barrel. (*Warehouse Collection*)

equipment and a stowage basket, and the Mk 10/2 was equipped with the ranging machine gun.

The Mk 11 was an up-armoured Mk 6, but equipped with infrared lights, a ranging machine gun and the stowage basket on the turret rear; the Mk 12 (FV4007) was similar but was developed from the Mk 9; and, finally, the Mk 13, the last 'official' British Army variant, was a Mk 10 equipped with infrared lights, a ranging machine gun and the rear stowage basket.

Chapter Two

Centurion Production and Sales

Although the majority of Centurions were constructed by Vickers-Armstrongs at Elswick, Newcastle-upon-Tyne, and at the Royal Ordnance Factory (ROF) Leeds, the vehicle was actually assembled at six separate plants.

AEC produced the first complete hulls, but subsequently the company had no further involvement in the project, and the Mk 1 was produced at ROF Woolwich, Leeds and Nottingham. The Mks 2, 3 and 5 were constructed by ROF Leeds, ROF Dalmuir (at Clydebank in Scotland), and at the Vickers-Armstrongs (later Vickers Defence Systems) factory at Elswick. The Mks 7 and 8 were produced by ROF Leeds and Leyland Motors, the latter operating from a new Ministry of Supply factory in Preston, Lancashire. Finally, the Mk 10 was constructed at ROF Leeds.

Production was in full swing by about 1950, with the tanks being produced at a rate of twenty or so a month. Peak production was achieved within another three years, in 1953, by which time the numbers had risen to about forty-five Centurions a month, with each requiring a production lead time of around thirteen weeks. As the vehicles were completed, the FVRDE specification decreed that every example would be subject to a routine inspection of components and a series of acceptance tests, both on the road and across country, to ensure that it had been manufactured in accordance with the specification.

As regards price, the development cost for the tank was said to be in the order of £5 million, and the construction cost per vehicle, in 1951, was £31–38,000 – somewhere around £1.025 million at 2011 values (by comparison, a Challenger 2 main battle tank cost £4.095 million in 1988). The cost of a spare gun barrel was around £9,000.

Other companies involved in the manufacturing process included David Brown, who manufactured the Merritt-Brown controlled-differential transmission, Vickers-Armstrongs and the Royal Ordnance Factories at Cardiff and Nottingham, who made the main guns, and the Rover Company and Morris Motors, who supplied the main and auxiliary engines; although the Meteor engine had been designed by Rolls-Royce

(*Above*) Constructed at the Royal Ordnance Factory Woolwich, and seen here parked outside the so-called 'shell foundry', T351701 was the second of a hundred Centurion Mk 1 A41* variants to be built; others came from ROF Nottingham and Leeds. The 17-pounder (76.2mm) gun is mounted in a rolled-steel welded turret as with the prototypes; note the circular escape hatch. (*Warehouse Collection*)

(*Opposite, top*) From the Centurion Mk 2 onwards, the welded turret was replaced by the cast design shown here. This photograph, which was taken at one of the Royal Ordnance factories in 1948, shows the turret, complete with gun, being lowered into position on the near-completed hull. The lower brackets for the removable side skirts can be seen between the bogies. (*Warehouse Collection*)

(*Opposite, below*) Brand-new Centurion Mk 3s in store awaiting issue to units. These tanks are fitted with the so-called 'B barrel' 20-pounder (83.4mm) gun which was fitted with a fume extractor. (*RAWHS*)

Centurion Mk 3 of the Canadian Army photographed at Worthington Tank Museum at CFB Borden in Ontario, Canada. Canada ordered 420 Centurions, with the British government agreeing to supply 205 during the period 1951–54. The total number supplied to Canada eventually reached 374. (*Balcer*)

and is almost always referred to as a Rolls-Royce product, development, design rights and manufacture of the engine had actually been transferred to the Rover Company in 1943, in what, these days, we would call a technology-transfer deal. Sub-assemblies were supplied by various external contractors as well as by the Royal Ordnance Factories at Patricroft (Manchester), Ellesmere Port (Liverpool) and Radcliffe (Wigan). Most electrical equipment was supplied by Lucas-CAV and Simms, and the gyroscopes and servo system that allowed the Centurion to fire on the move with a degree of accuracy was manufactured by the Manchester-based heavy electrical engineering company Metropolitan-Vickers (Metrovick).

Production finally came to an end in March 1962. The last gun tank, which was a Mk 10 numbered 03DA03, was constructed at ROF Leeds under contract KL/A/0147. One must assume that, had the demand existed, the production line at Leeds or at Vickers-Armstrongs would have been reopened because, even four years later, Centurion bridgelayers, ARKs and AVREs were still being shown to overseas buyers at the bi-annual exhibitions of British military vehicles held at FVRDE's Chertsey site.

The exact total of Centurions produced was 4,423 over an eighteen-year period (1945 to 1962), and the most numerous variant in production was the Mk 3, with a total of 2,833 produced. The major manufacturers were Vickers-Armstrongs, who constructed a total of 1,437 gun tanks, and ROF Leeds with a total of 2,392. Of the total number constructed, some 2,500 went for export, with sales worth more than £200 million, leaving somewhere around 2,000 for the British Army, although these were never all in service at the same time. However, compare this figure with the current numbers of British Army Challenger 2 main battle tanks: the total number ordered was 386 and the number anticipated to be in service after the 2010 Strategic Defence and Security Review (SDSR) was around 200.

Although the largest number of the Centurions went to the British Army, many were also exported, at first to other NATO countries. In December 1951 Prime Minister Winston Churchill, who had won the second post-war election in October 1951, forwarded a memo to the Secretary of State for War, Anthony Head, which suggested that the Americans would be prepared to buy Centurions with which to equip other NATO countries as well as possibly using the tanks themselves. In replying, the Secretary of State pointed out that the previous (Labour) administration had received a request from Canada for 420 Centurions, and had agreed to supply 205 tanks during the period 1951–54 at a price of around $187,000 each. This placed something of a burden on the manufacturing facilities, but nevertheless the Ministry of Supply, at the time led by Duncan Sandys, was asked to investigate the possibility of meeting the Americans' request and selling Centurions to the US government. On the American side, William L. Blatt indicated that the Harry Truman government would be prepared to purchase at least 500 and possibly 1,000 Centurions under the Mutual Defense Assistance Program (MDAP) that had been set up under the 'Mutual Defense Assistance Act 1949', to 'assist with Britain's balance of payments position'.

The price per tank was quoted at $210,000, which works out at around £74,000 ... at a time when the construction cost was less than £40,000! The terms of the 'Mutual Defense Assistance Act' required that any such sales made to the US government must be additional to the rearmament programme of the country concerned, but the Minister believed that an additional 750 Centurions could be produced during the period 1952–54 without impacting negatively on the British requirement for 2,500 tanks, providing that the US government supplied 'certain machine tools, materials and components'. A secret memo of 25 January 1952 stated that the matter was of some urgency and that unless firm proposals were put forward within seven days, the US government might well decide to drop the project altogether.

Clearly, some agreement was reached within a reasonable timescale ... and

Privately owned ex-Swiss Army Centurion Mk 5 or Mk 7, one of 200 supplied by Vickers-Armstrongs during 1955 and 1957 (designated Pz 55 or Pz 57); the tanks were eventually fitted with the 105mm L7 gun in 1960 and the designations were amended to become Pz 55/60 and Pz 57/60. Note the lighting equipment, which was altered to comply with Swiss road traffic legislation, and the spare road wheel carried on the nose. (*Simon Thomson*)

Canada was not the only Commonwealth country to take delivery of Centurions ... they were also supplied to Australia (11,144), New Zealand (12), India (245) and South Africa (541). This example stands outside the Queen Elizabeth II Army Memorial Museum at Waiouru in New Zealand. (*Winston Wolfe*)

A well-sheeted Centurion Mk 3 loaded onto the standard FV3601 50-ton drawbar trailer. The sheet is marked 'ROF Leeds', which would suggest that the tank is on its way from the factory to a storage depot, but the fact that the trailer has seemingly been abandoned in the middle of a residential street would indicate some sort of mechanical difficulty. (*CVRTC*)

between 1952 and 1954 some 216 Centurion Mk 3s, including a number of ARVs, together with spare parts and ammunition valued at $140 million, were supplied to the Royal Danish Army under the MDAP arrangements. (*See colour plate 4.*) Similarly, between 1953 and 1956 the Royal Netherlands Army received 591 Centurion gun tanks and forty-four ARVs under the same MDAP scheme. In 1957 a further seventy Mk 7 Centurions were purchased for the Royal Netherlands Army, but they were disposed of just six years later, apparently because the UNF thread forms led to an unacceptable degree of incompatibility with existing Dutch equipment! In 1958 the US government indicated that it wanted to purchase 105 up-gunning kits from Britain for the Centurions that were in service with Denmark and the Netherlands but there was something of an unseemly scramble for priority, with the Director, Royal Armoured Corps (DRAC) indicating that 'foreign armies' should not be given precedence over the domestic force.

As was the case with all MDAP materiel, ownership of the Centurions continued to reside with the US government and, in theory, the tanks were either returned to the US government or disposed of through the US Army's equipment disposals office at Wiesbaden. Of course, these MDAP vehicles were not the only Centurions to be sold overseas but others were supplied under straightforward contract arrangements, negotiated directly between the British government and the nation in question.

Chapter Three

The Centurion Crew

During most of the Second World War British tanks were operated by a crew of five but, with the deletion of the hull machine gun, the standard crew for the Centurion was four men, comprising the commander, gunner, driver and loader/operator, and for much of its life the Centurion was operated by conscripts, the last of whom did not leave service until 1963. As well as operating the tank, the crew was expected to check the operation of systems such as the turret traverse and gun elevation, fire extinguishers, radio sets, intercommunications, etc., at the 'first parade' of the day, to check that stowed items were secure, and to carry out a range of maintenance tasks at 'halts on the march' and at the end of the day. These tasks included checking the track tension and the condition of road-wheel tyres; checking and, if necessary, replenishing oil and water levels; checking fuel levels; cleaning out air filters; checking the condition of all drive belts; ensuring that the head and tail lamps and all warning lights were operating correctly; cleaning and oiling the guns, etc., etc. As you might expect, there was also a fair amount of record-keeping involved with regard to the specified maintenance tasks and the number of rounds fired by the main and auxiliary weapons.

The commander was the senior crew member. He was seated in the turret basket, on the right towards the rear, and had a second smaller folding seat on the turret wall. His role was to command the tank and the crew, and to receive and execute orders from higher command. He was also responsible for map reading and for directing the driver, and for ensuring that crew maintenance duties were carried out at the specified intervals.

The role of the gunner, who was located ahead of the commander, with the gun controls in front of him and to the right, was to aim and fire the main and coaxial guns when ordered to do so by the commander. He was also responsible for carrying out maintenance tasks related to the guns and the sighting equipment.

The driver was seated low down on the right of the hull, protected by the thick, sloping glacis plates. His seat was adjustable, and could be raised to allow the tank to be driven either fully closed-down or with his head out through the hatch. Of course, he was responsible for actually driving the tank in response to orders from the

(*Above*) With an instructor perched alongside the open hatch, a Junior Leader is instructed in the finer points of conducting a Centurion across country. The vehicle is a Mk 3 constructed by one of the Royal Ordnance factories and the photograph was taken in August 1965. (*Warehouse Collection*)

(*Opposite, top*) This Centurion is travelling across fresh desert scrub at about 10–15mph and is raising something of a dust cloud. The engine covers seem to be raised, perhaps to improve cooling, and the three men seated on the turret are certainly not regular crew members, which may indicate that the tank is undertaking some sort of trial. (*Warehouse Collection*)

(*Opposite, below*) A REME forward repair team attending to a Leyland-built Centurion Mk 3 during an exercise in BAOR. One team is removing the defective engine pack, a second man is repairing damaged track guards, while a third changes the radio set. (*Warehouse Collection*)

commander but his training, a process that incidentally required a course of up to four weeks' duration, included not only how to operate the tank and the driving controls, in terms of changing gear, steering, etc., but also how to choose concealed routes, how to negotiate natural obstacles, and how to select suitable firing positions. With a view to achieving maximum reliability, none of the controls was power-assisted and the Centurion was heavy and not always easy to drive, and called for considerable skill and physical strength ... some contemporary commentators pointed out that two hands were often required to change gear! The driver was also responsible for the day-to-day maintenance tasks associated with the engine, transmission and suspension, and, if one was fitted, with the mono-wheel fuel trailer.

Finally, the fourth crew member was described as the loader/operator. His main role was to load the main and coaxial guns, and also, before its use was discontinued – since it provided a constant hazard inside the turret on which the loader would often hit his head – to attend to the roof-mounted smoke-bomb thrower. The loader also doubled as the radio operator and was provided with twin radio sets located in the turret bustle. And, as if this wasn't enough, he was also responsible for storing ammunition, clearing jams in the weapons, ejecting spent shell and cartridge cases ... and for preparing tea and food for himself and the other crew members.

The crew requirements for the engineering tanks were similar, but obviously not identical. For example, although the Centurion ARV and the bridgelayer were also operated by a crew of four, their roles were decidedly different from those of the crews of the gun tanks: in the ARV, the crew comprised the commander, driver and two winch crew; and in the bridgelayer there was a commander, driver, co-driver and engineer. Both the BARV and the AVRE carried five crew, the former consisting of a commander, driver, radio operator and two recovery mechanics, and the latter a commander, driver, co-driver, gunner and loader/operator.

Chapter Four

The Centurion in Combat

The Centurion entered regular service with the British Army in December 1946, when a small number of Mk 1s and 2s were delivered to the 5th Royal Tank Regiment, 7th Armoured Brigade, which at the time was based at Hamm in Germany. By the end of 1948 the new tank was also in the hands of the other two regiments of 7th Armoured Brigade, the 1st Royal Tank Regiment and the 5th Royal Inniskilling Dragoon Guards. Ultimately, as well as being based in Britain, there were Centurions with British armoured regiments in Aden, Hong Kong and West Germany. An armoured regiment of the day was generally equipped with forty-eight Centurions in three squadrons of fifteen, with the remaining three assigned to the headquarters; a squadron consisted of four tank troops and a headquarters, each with three tanks, and each squadron also normally included a 'dozer tank in its complement. From the mid-1950s, in West Germany, six of the Centurions were replaced by Conquerors.

In the gun tank role, the Centurion enjoyed a more than twenty-year career with the British Army, but by the early 1970s most had been replaced in service by the new Chieftain, albeit some Centurions were retained for driver training. It was not quite the same story for the engineer variants, and both ARVs and BARVs, as well as 'dozer tanks, AVREs, bridgelayers and ARKs, remained in service into the 1970s and, in some cases, well beyond ... astonishingly, a small number of AVREs actually saw active service in the Gulf War in 1990.

The Centurion arrived too late to see action during the Second World War but nevertheless many of the tanks still spent their working lives in Germany, where they were assigned to the 1st, 3rd and 4th Armoured Divisions of BAOR until their replacement by Chieftains, a process which began in November 1966. Fortunately, there was no live action in Europe, and instead the tanks of BAOR spent their working lives endlessly training for a Soviet invasion that never came; the first significant BAOR exercises in which Centurions were involved were Operation Broadside 1 and Operation Broadside 2. Involving 7th Armoured Division and 2nd Infantry Division, the exercises were carried out in late September 1950, and were intended to 'practise movement and concentration in the face of enemy air superiority, and to carry out operations on wider fronts entailing movement laterally

All Centurion Mk 3s were armed with the 20-pounder (83.4mm) gun. This one was photographed in front of the ancient Roman amphitheatre of Sabratha in the Libyan city of Tripoli. The markings on the lower glacis plate indicate that the tank belongs to the British 1st Armoured Division, reformed in 1960 after being disbanded in 1945. (*Warehouse Collection*)

and from front to rear, and quick concentration for attack and dispersion afterwards'. Little was said of the performance of the Centurions, but it was stated that the length of the barrel made the tank difficult to conceal.

The tank saw its first combat during the Korean War, with a number of Mk 3s, originally destined for Australia, being diverted to the 8th King's Royal Irish Hussars – generally simply described as the 8th Hussars – towards the end of 1950, where they joined US Army M26 Pershings facing the Chinese and the North Korean People's Army (NKPA) at New Year 1951. Centurions went into action at the Battle of Imjin River in April of that year, where they were famously used to provide cover for the withdrawing infantry of the 29th Brigade. By May 1951 the British Army had sixty-four Centurions in Korea and by the end of the year, when the Hussars were relieved by the 5th Royal Inniskilling Dragoon Guards, the Centurions were dug in amongst infantry positions on high ground facing the enemy. There was little movement as the British Centurions and the Chinese and North Korean T-34/85s exchanged fire across no-man's-land. During the following year the tanks were involved in limited armoured raids across the unfavourable terrain, some of which took place in sub-zero temperatures. In late 1952, with the war grinding on and neither side able to gain the upper hand, the 5th Royal Inniskilling Dragoon Guards were relieved by the 1st Royal Tank Regiment, and the Centurions played a significant role in repelling Chinese forces during the second Battle of the Hook in 1953. During one night's action 504 high-explosive (HE) 20-pounder rounds were rained down on the enemy.

Centurion ARVs were also first deployed in Korea, replacing older vehicles based on the Churchill infantry tank. Their performance was described as 'excellent'.

The Centurion was highly praised for its all-round performance, and particularly for its apparent ability to go anywhere, while the minimum elevation (-10 degrees) of the main gun allowed the Centurion to operate almost completely concealed in a 'hull down' position. Its lack of vulnerability under fire provided a real boost to the morale of the fighting men ... one official report specifically singled out the lack of internal effect from a hollow charge from a 3.7in Russian bazooka or a captured US Army recoilless rifle which created a 3in deep hole in the back of the turret but failed to penetrate. Several tanks also received multiple direct hits that caused little damage to the tanks and no injuries to the crews ... there is a story of two Centurions that had to be abandoned in Korea, with unsuccessful attempts being made to destroy them using 20-pounder armour-piercing shot to prevent them falling into enemy hands; undestroyed, they were eventually recovered, more or less intact. Another story described a Centurion in the 29th Brigade's sector during March 1952, sliding sideways from the top of a razor-edged ridge, gathering speed down the slope as the tracks failed to grip the frozen ground. Eventually the tank

At the time when the Centurion started to enter service, the standard British Army tank transporter was the American Diamond T Model 980/981 used in conjunction with a 40-ton drawbar trailer. The Diamond Ts shown here, with their Centurion loads, belong to 312 Tank Transporter Squadron, formed in 1951 and based at Fallingbostel in Germany. With a combat weight approaching 50 tons, the Centurion is certainly overloading the trailer. (*Warehouse Collection*)

A Leyland-built Centurion Mk 3 photographed during a Royal Armoured Corps (RAC) demonstration in 1960. The turret is traversed to the rear. (*Warehouse Collection*)

From about 1951 the Diamond T tank transporters started to be superseded by the 50-ton Thornycroft Antar, seen here in its original Mk I ballast-bodied form. The Centurion, a Mk 3 from 1948, is being winched onto the trailer. (*Warehouse Collection*)

In 1958 the Antar Mk 2 was replaced by the much-improved diesel-engined Mk 3 (seen here) and Mk 3A, the latter designed for use with a drawbar trailer. Rated at 60 tons, the Antar Mk 3 was able to cope with both the Centurion and the heavier Chieftain that followed. Note the FV1801 Austin Champ in attendance. (*BCVMT*)

somersaulted three times, landing in a minefield, in which it caused considerable mayhem, before arriving at the bottom, on its tracks, but with the gun barrel severely bent and the turret off its ring – indeed, as the contemporary report put it, the 'tank generally was in considerable confusion'. All of this happened in full view of the enemy! The crew was shaken and embarrassed but generally uninjured and the tank was abandoned. However, this wasn't the end of the sorry saga and it was decided that recovery was impractical since it would require two ARVs, one of which would have to be lowered down by the other. The Royal Engineers were asked to destroy the gun-stabiliser equipment to prevent it falling into enemy hands but, overestimating the size of charge required, managed to 'set everything off'. The report of the incident ended by stating that the 'tank will be of no value to the enemy'!

Hostilities in Korea came to an end on 27 July 1953 with no proper resolution of the conflict. All things considered, the Centurion was an extremely capable machine, able to fire accurately even at its maximum range, and able to traverse the most rugged and challenging terrain ... even in the stickiest mud, the tracks never sank more than 12in into the ground and cross-country performance was always considered to be excellent. The extreme cold of the Korean winter sometimes caused problems with tracks failing to gain traction on the frozen ground and, on occasions, the track brakes, which were entirely mechanical, became inoperative due to icing leading to at least one runaway. However, the Centurions had proved themselves to be worthy opponents.

Little more than three years later, British Army Centurions of the 1st and 6th Royal Tank Regiments were deployed to Egypt in November 1956, as part of the joint Anglo-French-Israeli operation intended to wrest back control of the Suez Canal Zone out of the hands of President Nasser. Despite a shortage of landing craft, which restricted the number of vehicles available to ninety-three, Centurions were successfully landed and fought alongside the French AMX-13 light tanks, capturing Port Said in November, before the governments involved finally bowed to UN pressure and withdrew the troops on 23 December.

In September 1960, along with 80,000 soldiers from four nations, Centurions took part in the largest land, sea and air exercise staged in the northern Federal Republic of Germany (West Germany) during the Cold War era. Dubbed Operation Holdfast, the exercise was designed to test the effectiveness of NATO defences in the Jutland peninsula. Tanks of the 5th Royal Inniskilling Dragoon Guards were put ashore from German and British landing craft in Eckenforde Bay in Schleswig-Holstein, Germany, while others were brought up by transporter to a holding area south of the Hamburg Lubbeck autobahn. The defenders made mock nuclear strikes against the attacking troops but British armour, forming part of the

A heavily ballasted Centurion Mk 3 being loaded onto an eight-wheeled Cranes 50/60-ton tank-transporter semi-trailer. The trailer is coupled to a Rolls-Royce-engined Scammell Constructor 6x6 prime mover of a type that was offered for export by the company for tank-transporting duties. (*Scammell Lorries Limited*)

A 20-pounder (83.4mm) gun-equipped Centurion of the 6th Royal Tank Regiment painted in desert sand, almost certainly for use in the unfortunate Suez affair. The truck is a 1-ton Humber of the FV1600 series. (*Warehouse Collection*)

attacking force, pushed inland to reach within 2 miles of the crucial Kiel Canal. At the end of the nine-day exercise the conclusion was that NATO was well prepared to withstand such an assault from the joint forces of the Warsaw Pact ... and the Centurion was seen as a valuable element of the exercise.

Centurions of the Royal Scots Greys were also deployed in Aden (Radfan) during the 1963/64 uprising against British control.

Although most of the Centurion gun tanks had been replaced within twenty years, it was not quite the same for the engineer variants and both ARVs and BARVs, as well as 'dozer tanks, AVREs, bridgelayers and ARKs, remained in service into the 1970s and beyond. In July 1972 four Centurion AVREs of 26 Armoured Engineer Regiment were deployed to Northern Ireland aboard HMS *Fearless* and were used to clear Republican roadblocks that had been erected around the Rossville Flats on the Creggan Estate in Derry-Londonderry. The roadblocks were believed to be booby-trapped. Described as Operation Motorman, the exercise was considered to be extremely sensitive for obvious reasons and was conducted during the early hours of the morning; in order to minimise the danger of sensationalist headlines, the tanks were operated with the guns covered and traversed to the rear.

In 1982 a pair of surviving Centurion BARVs were operated from the two LPD ('landing platform, dock') vessels HMS *Fearless* and HMS *Intrepid* during the Falklands War. One broke a drive chain and remained unused for most of the conflict. More recently, both of these BARVs were also deployed to Iraq, fighting in both Gulf Wars before being retired from service in 2002; at least one has survived in private hands.

Astonishingly, a small number of ARVs and AVREs, fitted with additional passive and explosive reactive armour (ERA), saw active service during Operation Granby, the British contribution to the liberation of Kuwait during 1990/91. Despite the Ministry of Defence (MoD) having lost a large percentage of the remaining Centurion spares in a fire at the Donnington stores in 1988, AVREs of 32 Armoured Engineer Regiment played their part in helping to move some 850,000 tons of earth, and in blasting through Iraqi defences. Some spares support was drawn from Aviation Jersey Ltd, who had acquired the entire stock of Centurion parts from the Netherlands in 1990.

In 1968 and 1969 Royal Australian Armoured Corps Centurions, including tank 'dozer variants, saw action in Vietnam, where most were equipped with 100-gallon auxiliary fuel tanks attached to the rear of the hull, giving the designation Centurion Mk 5/1 (Aust). The decision to send Australian Centurions, consisting of more than twenty gun tanks, two bridgelayers, two 'dozers and two armoured recovery vehicles, had been taken in October 1967 as part of Australia's increasing involvement in this ugly conflict. Operating under US command, the Centurions were involved in the Tet Offensive in January 1968, and during their service in south-

A Centurion armoured recovery vehicle (ARV, FV4006) photographed during the liberation of Kuwait in 1990/91. The registration number (00ZR48) indicates that this vehicle was converted from a Mk 1 or Mk 2 Centurion gun tank dating from the immediate post-war years. Note the additional composite armour applied to the sides of the vehicle in the form of panels. (*Tank Museum*)

Centurion AVREs (FV4003) also took part in the liberation of Kuwait in 1990/91. This example carries additional armour on the turret face and the flanks of the vehicle, and lacks the cradle normally used for carrying the fascine bundle. (*Warehouse Collection*)

east Asia they acquitted themselves well in the difficult terrain, including rice paddy and jungle. The Centurions were returned to Australia at the end of 1971.

Australian Centurions were never involved in combat elsewhere. However, back in October 1953 the British and Australian armies had exposed what has been described as a 'near brand-new' Leeds-built Centurion Mk 3 of the Australian 1st Armoured Regiment (06BA16, Australian Army number 169041) to a nuclear blast test at Emu Field as part of Operation Totem One. The tank was parked less than 500 yards from the epicentre of the blast, with its engine running. Although it had run out of fuel by the time the test was concluded, and had sustained minor damage, for example to antenna and stowage bins, once it had been decontaminated it was capable of being driven from the site. The tank was subsequently repaired and used in the Vietnam War; in May 1969, during a fierce engagement with the enemy, it was penetrated by a rocket-propelled grenade (RPG) that wounded all of the crew in the turret. The RPG entered the lower left side of the fighting compartment, travelled diagonally across the floor and came to rest in the rear right corner. 169041 was given its third base overhaul in 1970, spending some time in storage before being reissued to the 1st Armoured Regiment. By 1976 Centurions had been phased out of Australian service, having been replaced by the West German Leopard, but 169041 survived and is currently located at Robertson Barracks in Australia's Northern Territory, where it has been restored to running condition. Nicknamed 'the atomic tank', it is occasionally brought out for ceremonial duties.

Three more Centurions were involved in nuclear testing in Australia during the British government's Operation Buffalo tests held at Maralinga in September/October 1956. The operation involved the detonation of four separate nuclear devices, code-named One Tree, Marcoo, Kite and Breakaway, two of which (One Tree, with a yield of 12.9 kilotons, and Breakaway at 10.8 kilotons) were Red Beard tactical bombs exploded from towers, while Marcoo (1.4 kilotons) and Kite (2.9 kilotons) were Blue Danube bombs, the first exploded at ground level, and the second released by an RAF Valiant bomber from a height of 35,000ft. This was the first aircraft launching of a British atomic weapon. During one of these trials the three Centurion Mk 3s were placed at roughly 440 yards, 880 yards and 1760 yards from ground zero. Even at 440 yards the blast damage was only superficial, being largely confined to the external sheet metal, and was not sufficiently serious to have prevented the vehicles fighting again. The report of the trial stated that the Centurion 'was capable of taking heavy punishment at the range, and with the weight of bomb used, without being disabled to a non-fighting state'. One vehicle (05BA60) was quickly made serviceable and drove some 80 miles after the blast; the gun was also test fired with no recorded loss of accuracy. The fate that might have befallen the luckless crew had this not been an exercise was not recorded.

Egyptian Centurions saw action during the Six Day War with Israel in 1967, with most being captured by the Israeli Defence Force (IDF). Ironically, the Egyptian Army captured a similar number of 105mm-equipped Centurions from the IDF during the Yom Kippur War of October 1973, although by this time Egypt was operating predominantly with Soviet tanks and equipment.

In India Centurions were deployed during both of the border wars with Pakistan in 1965 and 1971. The 1965 war lasted five weeks and is considered by many to include the largest tank battle in military history since the Second World War. A total of 186 Indian Centurions fought alongside some 340 ageing M4 Sherman tanks of the Indian Army, with both Shermans and M47 and M48 Patton tanks of American origin opposing them on the Pakistani side. The Centurions proved themselves to be superior in most respects to the more modern (and more complex) American tanks, and were able to withstand 90mm armour-piercing shells fired from the powerful M63 guns of the Pattons.

Photographed in the Valley of Tears in the Northern Golan Heights, this 150mm L7-equipped Sho't Kal Centurion of the Israeli 7th Armoured Brigade provides a memorial to the Yom Kippur War of 1973. During this action 150 Israeli tanks faced more than 1,400 Syrian tanks across the Golan Heights. Although the Centurions fought well, this was to be their last combat before being replaced by the Israeli-designed Merkava. (Isrredlabel)

Jordanian Centurions went into action in 1970 to counter Syrian border incursions during the conflict with the Palestinian guerrilla organization Black September that ended in July 1971 with the expulsion of the PLO to Lebanon. In 1973 Jordanian Centurions were in action again in the Golan Heights.

South African Centurion-based Semel tanks were deployed in Namibia (South West Africa) against the military wing of SWAPO (South-West Africa Peoples Organisation) during the fight for independence that endured between 1966 and 1988. The more heavily modified Olifants were deployed against Angolan forces during the Angolan civil war in 1987, where they were fighting against Soviet-built T-34/85s and T-55s.

Outside of Britain, Israel was not only the largest user of Centurions, but was also the nation with the most experience of using the tank in combat. Although only

Mk 3 or Mk 5 Centurion, one of a large batch (650) produced by the Royal Ordnance Factories from 1950. The main gun is the 'B barrel' 20-pounder (83.4mm). (*Warehouse Collection*)

around 250 Centurions were supplied new to Israel, many more were acquired as surplus or were captured during various campaigns with Israel's neighbours: during the 1967 Six Day War, for example, Israel captured thirty Centurion tanks from Jordan. At one time the IDF was able to deploy a total of around 1,000 Centurions, some 25 per cent of the total production figure, all of which were eventually equipped with the 105mm L7 gun. Poor maintenance and abuse of the tanks in the Israeli deserts by the largely conscripted crews initially gave the Centurion a poor reputation. A company of Israeli Centurions was fired on by Syrian T-55 and T-62 tanks at Nukheila in 1964; despite firing some eighty-nine rounds of 105mm ammunition in an exchange that lasted ninety minutes, not one Syrian tank was hit. Things began to improve when General Israel Tal took command of the Israeli armoured corps towards the end of 1964, and standards of training, maintenance and discipline rose significantly. In a second border incident at Nukheila one Israeli Centurion destroyed two Syrian Panzer IVs. A year later Israeli Centurions destroyed Syrian earth-moving equipment that was being used to divert the Jordan

River. Centurions were among some 800 Israeli tanks successfully deployed against assembled Arab forces in the Six Day War of 1967, the tanks being called upon to fight again in the 1973 Arab-Israeli Yom Kippur War, where they were exploited to advantage in a hull-down position against the largely Soviet tanks of the opposing Arab forces. Although no longer deployed as gun tanks, small numbers of Israeli Centurions continue to survive, re-equipped as heavily armoured armoured engineers' vehicles, designated Puma, NagmaSho't, Nakpadon and Nagmachon.

Many of the Centurions sold to customers around the world saw no active service with their original owners, including those vehicles supplied to the Danish, Netherlands, Swedish and Swiss Armies. However, a few Centurions certainly remained in service into the 1990s, and many, including some British AVREs, were able to be maintained by virtue of the large strategic reserve of parts that had been purchased from the Netherlands government by Aviation Jersey Ltd on behalf of the NATO powers. In a huge operation, every case of parts was opened, quickly examined and then marked as fit for keeping or to be scrapped before being moved to the island of Jersey, where they were held for redistribution within NATO as required.

However, it is impossible to hold back the march of time indefinitely, and improvements in automotive performance, tank guns, and target acquisition and sighting equipment inevitably meant that the Centurion was effectively obsolete. No longer suitable for front-line service, those British Army Centurions that were not scrapped would have certainly suffered the ignominious fate of being used as range hard targets or being sold to more impecunious nations. It was the same story elsewhere, with many similarly superseded by more modern equipment and surplus vehicles sold to other nations. The situation in Denmark was typical. Many of the nation's 216 Centurions remained in service into the 1990s, serving alongside 120 Leopard 1A3 main battle tanks that had started to enter service in February 1976, but the Treaty on Conventional Armed Forces in Europe (CFE), signed in 1990 by NATO members and the Warsaw Pact nations, restricted Denmark to 300 main battle tanks and some 146 remaining Centurions were destroyed or taken out of commission between 1993 and 1995.

However, a handful of Centurions have survived in museums around the world, and the relatively low cost of surplus Centurions in the 1990s means that there are also more than a few in private hands.

Chapter Five

The Centurion Described

The Centurion was of conventional layout for a tank of the period, consisting of a hull of welded armoured steel plate, with a centrally placed rotating turret mounting the main gun together with a coaxial machine gun. Although the appearance was not especially distinctive, and the design must be considered unexceptional in most respects, in the interests of avoiding so-called 'friendly fire' incidents the Centurion's air recognition features provide a good description of the visual aspects of the design. These were said to include the small cupola offset to the right of the turret; pannier stowage bins on the turret sides; the position of the turret itself, which was well forward on the hull, with a distinctive rear overhang; the prominent gun mantlet; a heavily ribbed rear deck; the exhaust silencers on the rear track guards; and the track skirts.

In a major departure from the design of previous British tanks, the sides of the hull were described as being 'boat-shaped', meaning that the sides sloped inwards from the top to help deflect blast. The glacis plate was also sloped to increase the effective thickness. The main structure of the hull was assembled from ten steel plates, consisting of two side plates, the upper and lower glacis plates (the lower plate sometimes being described as the nose plate), two top plates, the floor and three bulkheads. The turret ring was welded to the side plates and to the two top plates. There were no drain points in the floor, drainage of the hull being effected by means of a bilge pump mounted in the engine compartment. Removable armoured skirts were attached to the sides, designed to protect the upper track run from infantry-launched shaped-charge high-explosive anti-tank (HEAT) projectiles. The hull was fitted with lifting eyes and lashing shackles at all four corners and a rotating tow hitch at the rear.

The turret of the Mk I was assembled from welded steel plates, but all other variants were fitted with a three-man cast-steel turret cast in one piece, with a separate roof plate welded into place from the outside; detail differences were apparent in the design of the turret throughout the life of the tank. When viewed from above, the turret was roughly circular in shape with a large counterweight to the rear, and a hinged ammunition re-supply hatch on the left-hand side. Early

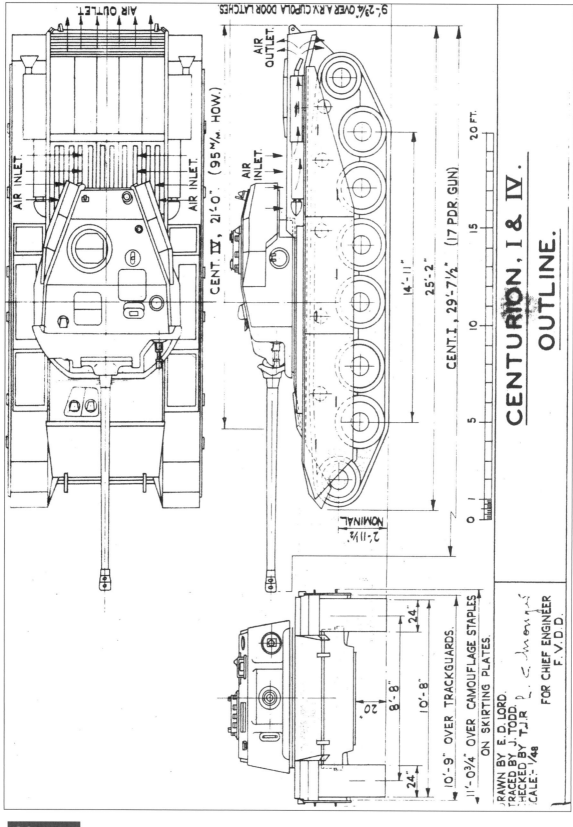

CENTURION, I & IV.
OUTLINE.

AIR OUTLET. AIR OUTLET.

9'-2⅜" OVER A.R.V. CUPOLA DOOR LATCHES.

AIR OUTLET.

AIR INLET. AIR INLET.

AIR INLET.

CENT. IV, 21'-0". (95 M/M. HOW.)

CENT. I, 29'-7½" (17 PDR. GUN)

14'-11"

25'-2"

20 FT.

15

10

5

0 1

NOMINAL 2'-11½"

24"

8'-8"

10'-8"

20"

24"

11'-0¾" OVER CAMOUFLAGE STAPLES ON SKIRTING PLATES.

10'-9" OVER TRACKGUARDS.

DRAWN BY E.D. LORD.
TRACED BY J. TODD.
CHECKED BY T.J.R.
SCALE:- 1/48

FOR CHIEF ENGINEER
F.V.D.D.

Dated June 1947 and drawn by E.D. Lord, FVDD (Fighting Vehicles Design Department) drawing TD36515 gives outline dimensions for the Centurion Mk I (and the Mk 4 had it gone into production). *(Warehouse Collection)*

Colour Profiles

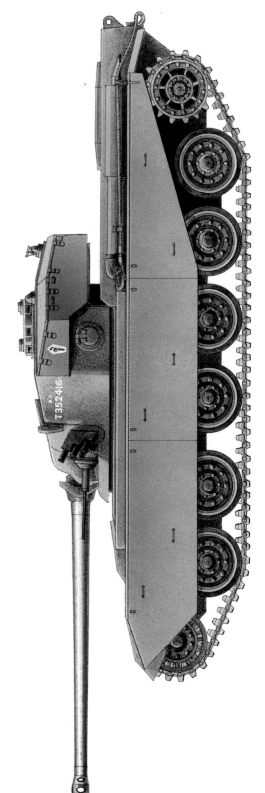

Plate 1a

Plates 1a and 1b: Centurion Mk 1 (A41*) gun tank; WO registration number T352416 (subsequently, 03ZR70)

The Centurion Mk 1 is equipped with the 17-pounder (76.2mm) gun in a rolled-steel turret. The tank is painted in the markings worn during the troop trials dubbed 'Operation Sentry', when it was attached to the 22nd Armoured Brigade, 5th Royal Inniskilling Dragoon Guards, before passing to 5th Royal Tank Regiment. The Mk 1 was constructed at the Royal Ordnance Factories at Leeds, Nottingham and Woolwich.

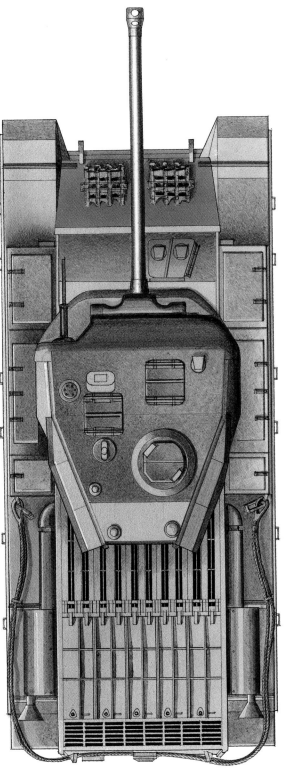

Plate 1b

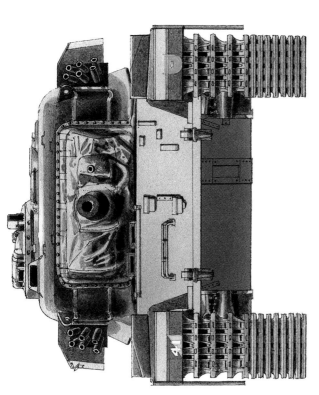

Plate 2a

Plates 2a and 2b: Centurion Mk 3 gun tank;
WO registration number 03ZR71
The tank is equipped with the 20-pounder
(83.4mm) gun in a cast turret. It is painted in
the markings of 3 Troop, C Squadron, 1st Royal
Tank Regiment, part of the Commonwealth
Division, during the Korean War. The Mk 3 was
constructed at the Royal Ordnance Factories at
Leeds and Dalmuir, and by Vickers.

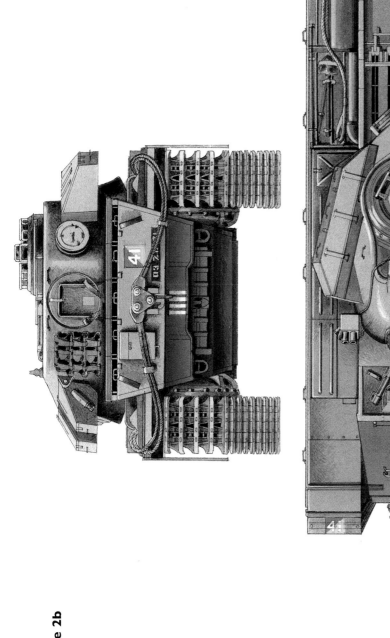

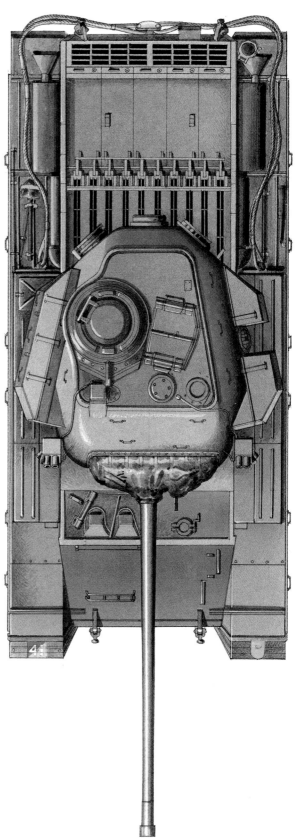

Plate 2b

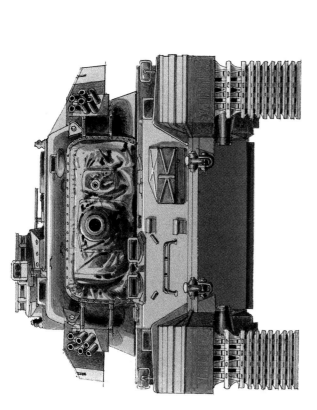

Plate 3a

Plates 3a and 3b: Centurion Mk 5/1 (FV4011) up-armoured gun tank

This variant is equipped with the 105mm L7 gun. The armoured side skirts ('bazooka plates') were often removed from tanks in service, particularly from those Centurions serving with the Royal Australian Armoured Corps. The Mk 5 was constructed at the Royal Ordnance Factories at Leeds and Dalmuir, and by Vickers.

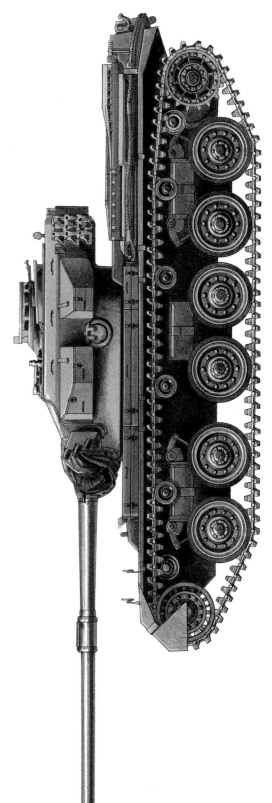

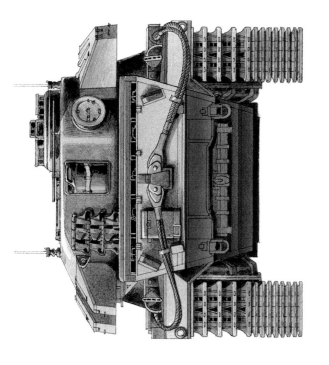

Plate 3b

Plate 4

Danish Army Centurion Mk 3 gun tank; modified to Mk 5 configuration

Equipped with the 20-pounder (83.4mm) gun; note the fume extractor indicating that the gun has the 'B' barrel. Danish Centurions were supplied under the US Government's Mutual Defense Assistance Program (MDAP); some were subsequently modified to Mk 5/2 configuration by fitting the 105mm L7 gun.

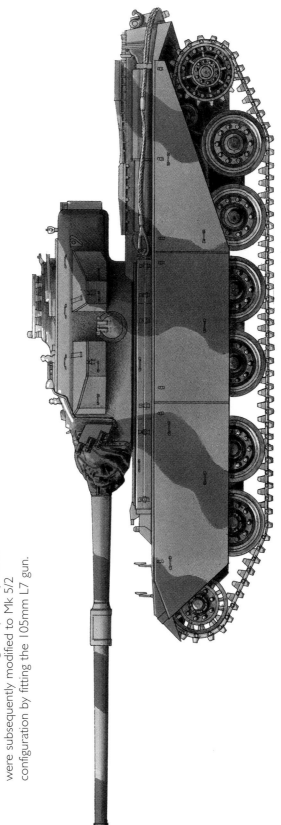

Plate 5

Swedish Army Centurion Mk 3 (*Stridsvagn 81*)
gun tank; Swedish registration number 80342
(WO registration number 05BA71)

This was one of the first six Centurion tanks
sold to Sweden in 1953 but was subsequently
upgraded to Mk 5 configuration (*Stridsvagn 102*)
by being fitted with the 105mm L7 gun.

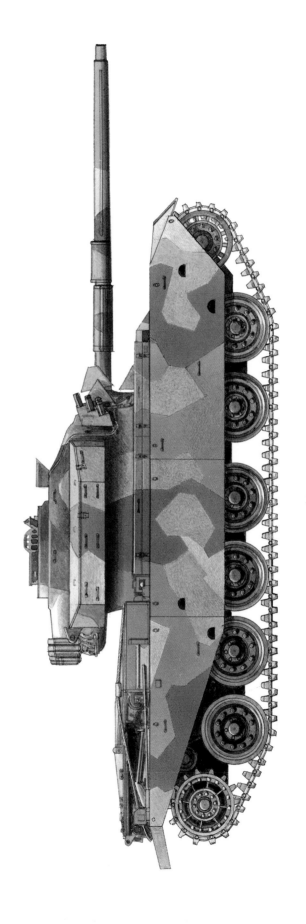

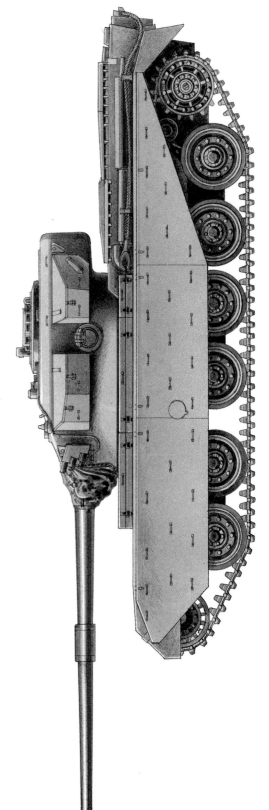

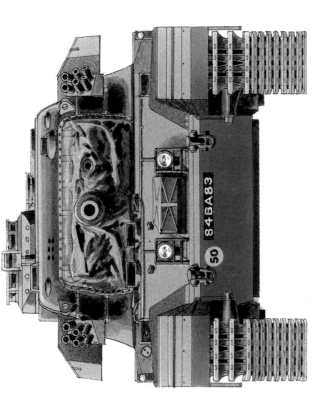

Plate 6a

Plates 6a and 6b: Centurion Mk 7 (FV4007) gun tank; WO registration number 84BA83 Equipped with the 20-pounder (83.4mm) gun with the 'B' barrel, the Mk 7 was redesigned by Leyland, and was constructed at the Royal Ordnance Factory at Leeds and at Leyland Motors, Preston.

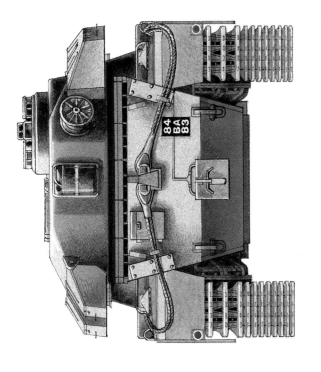

Plate 6b

Plate 7a

Plates 7a and 7b: Israeli Defence Force (IDF) Sho't Kal diesel-engined gun tank with 105mm L7 gun

The white chevron marking is typical of tanks that took part in the Yom Kippur War in October 1973. The hull has been modified at the rear to accept the larger Teledyne Continental diesel engine.

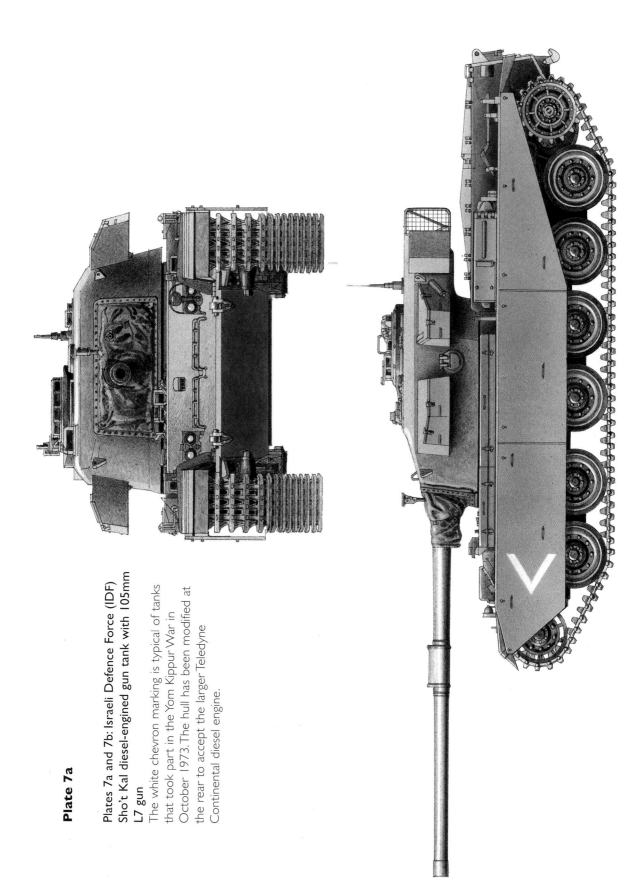

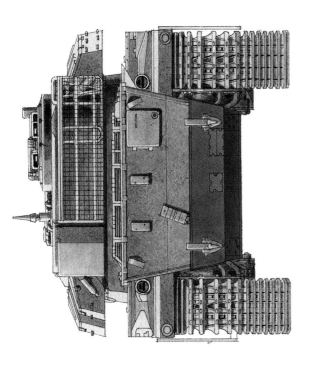

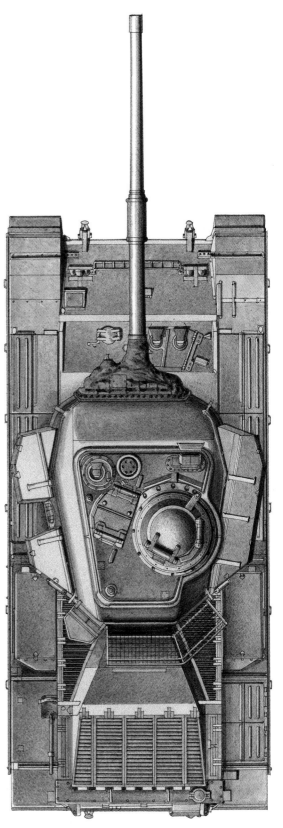

Plate 8a

Plates 8a and 8b: Centurion Mk 5 AVRE (FV4003)

Equipped with the 165mm L9 demolition gun, all the production AVREs (armoured vehicle, Royal Engineers) were converted from obsolete Mk 5 gun tanks. In service with the Royal Engineers, the AVRE saw action in the Gulf War in 1990.

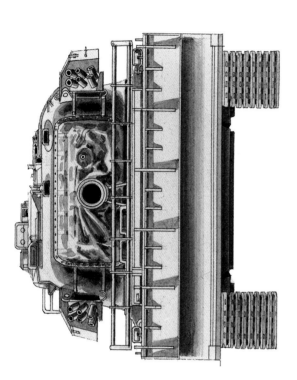

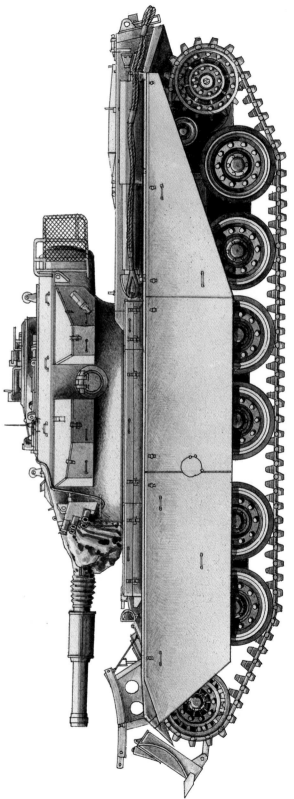

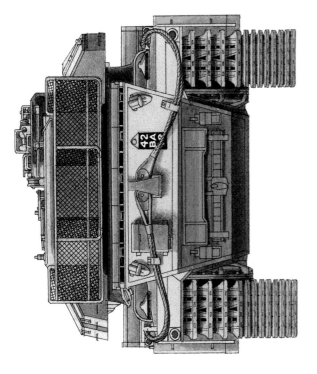

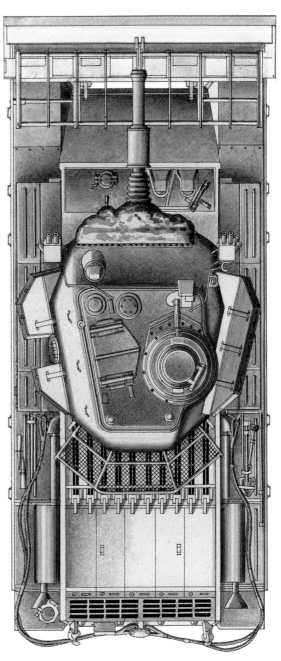

Plate 9

Centurion BARV (FV4108)

The beach armoured recovery vehicles (BARV) were based on the hulls of obsolete Mk 1, 2 and 5 gun tanks and were constructed by the Royal Ordnance Factory at Woolwich. In service with the Royal Electrical & Mechanical Engineers (REME).

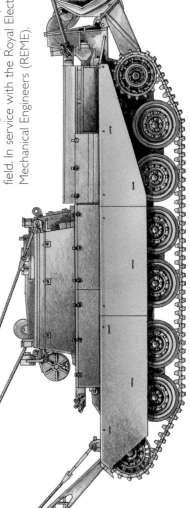

Plate 10

Centurion Mk 2 ARV (FV4006)

Armoured recovery vehicle (ARV) showing the aluminium lattice jib erected on the front of the hull for lifting turrets and power packs in the field. In service with the Royal Electrical & Mechanical Engineers (REME).

variants were equipped with a roof-mounted 2in smoke bomb thrower, which was designed to fire high explosive (HE), white phosphorous (WP) and illuminating flares out to a range of about 300 yards. This was subsequently deleted, but almost all variants were fitted with multi-barrelled smoke dischargers on the turret sides, each firing a maximum of six 'number 80 white phosphorous (WP) grenades' and providing 360-degree coverage to a range of about 60 yards. Originally, there was also a large rectangular access hatch at the rear of the turret, designed to allow the barrel of the 17-pounder gun to be withdrawn, but once the 17-pounder was superseded in production by the 20-pounder, the hatch was welded in place, before subsequently being eliminated altogether. The turret rotated on 164 32mm steel balls under electrical power, running on a 74in diameter flame-hardened turret ring; a full 360-degree rotation of the traversing system took 20–25 seconds. Large pannier bins were attached to the turret sides, originally with top-mounted access covers but in later production using side access covers. Lifting rings were welded to the upper corners of the turret to assist in removal.

According to the definitive *Janes Armour & Artillery*, no data has been officially released into the public domain on the thickness of armour of the Centurion. However, it is broadly accepted that, as originally envisaged, the maximum thickness of armour was as follows:

- 127mm on the gun mantlet
- 89mm on the turret sides
- 29mm on the turret roof
- 89mm at the turret rear
- 76mm (57mm for A41), measured at 57 degrees on the upper glacis plate, and at 45 degrees on the lower plate
- 51mm at 21 degrees on the hull sides
- 32mm at the hull rear
- 17mm on the hull floor

By the time the Mk 3 entered production in 1947/48, the maximum thickness of the gun mantlet had been increased to 152mm, and the thickness of the turret sides to 89mm. From the Mk 8/1 onwards the thickness of the glacis plate was also increased, to 127mm at 57 degrees (upper) and 45 degrees (lower).

Designed for what had become the standard British tank crew of four men, the hull was divided into three compartments. At the front was the driving compartment, with the driver entering the tank via an access hatch in the front of the hull that consisted of two spring-assisted side-hinged interlocked covers, the right-hand of which was fitted with a pair of periscopes. To left of the driver, and protected by a longitudinal steel divider, was an ammunition stowage bin, a 10-gallon container for fresh water, a pair of carbon-dioxide (CO_2) cylinders for the automatic

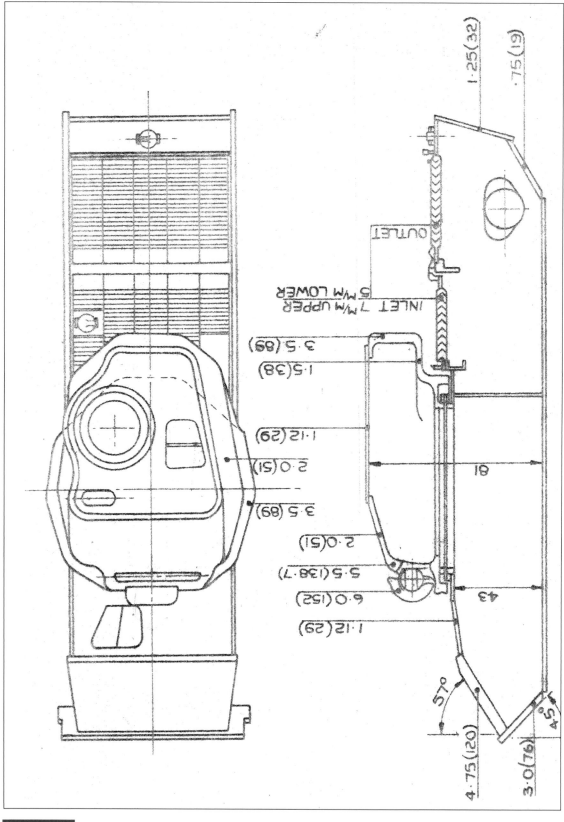

FVRDE diagram (drawing FV23192I, sheet 3) showing the thickness of armour at various points for the hull of the Mks 7, 8, 9 and 10. The drawing is dated April 1956. (*Warehouse Collection*)

fire-extinguishing system, and the four large 6V metal-clad vehicle batteries; in the Mks 7 and 8, the fresh water tanks were in the fighting compartment rather than alongside the driver. The driver's controls consisted of conventional clutch, brake and accelerator pedals, together with a centrally positioned gearshift control and a pair of levers coupled to the steering brakes; there was a handbrake lever positioned to the extreme right, rather too close to the right-hand steering lever. The instrument and switch panels were placed above the handbrake lever, with the former angled towards the driver.

In the fighting compartment, located in the centre of the hull, there was accommodation in the turret basket for the commander and the gunner. The commander's cupola, positioned on the right-hand side of the turret roof, incorporated a 22in diameter access hatch, and was provided with episcopes to ensure all-round vision. The fourth crew member was the loader, but he also doubled as the radio operator, with the selection of appropriate radio equipment made according to the role of the tank – originally these would have been number 19, 31, 38 and 88 sets of the Second World War period, the first of these also providing intercom facilities for the crew. Once the Larkspur VHF communications equipment started to be introduced in the mid-1950s, the installation would have comprised either B47 and C42 sets, or B47, C42 and C45 sets. An infantry telephone was fitted to the rear of the hull, allowing men on the ground to communicate with the tank crew by connecting into the intercom system.

An electric ventilator installed in the turret roof provided a positive pressure inside the hull and kept the crew compartment reasonably clear of heat and fumes, and there was provision for installing a heating system. There was no NBC (nuclear-biological-chemical) filtration system, but an automatic fire-extinguishing system was installed inside the engine compartment operated by five flame switches, two over the fuel tanks, two over the radiator cowls and one over the gearbox. In the event of the system being triggered, gaseous carbon-dioxide fire extinguishant was drawn from a pair of 7lb bottles in the driver's compartment and dispersed through five nozzles in the engine compartment. A warning horn sounded when the automatic system was tripped, and test buttons allowed the system to be checked manually. There were also two methylbromide and two tetrachloride manual fire extinguishers for use by the crew.

To the rear, separated from the fighting compartment by a fireproof steel bulkhead, were the main and auxiliary engines, the generator, the transmission, the huge twin air cleaners and the water/glycol cooling system. A large access plate in the bulkhead could be removed to allow fitters into the engine compartment for maintenance. The engine was installed along the axis of the tank, facing to the rear, with the gearbox arranged transversely at the very back, and with the hinged radiators of the cooling system placed on top of the gearbox.

All Centurions were fitted with a version of the Rolls-Royce (Rover) Meteor Mk 4 engine, a V12 unit based on the iconic Merlin aircraft engine of the Second World War, and producing around 550–650bhp from 27 litres. The photograph shows the Meteor Mk 4B that was fitted to Centurions Mk 3 to Mk 10, with a power output of 620–630bhp. (*Rolls-Royce Limited*)

In all variants the engine was the Rover-produced Rolls-Royce-designed Meteor Mk 4. Based on the iconic Merlin aircraft engine, the Meteor was a formidable naturally aspirated V12 petrol engine with a displacement a fraction over 27 litres. Five different versions of the Meteor engine were used, according to the particular variant of tank, with the power output ranging from 550bhp to 650bhp, according to the variant. The prototype Centurions and very early examples of the Mk 1 were fitted with the Meteor Mk 4, with a compression ratio of 6:1 giving a power output of 550bhp at 2400rpm. Centurions Mk 1s and 2s, from tank number 101, were fitted with the Mk 4A engine, in which the compression ratio was raised to 7:1, bringing the power output up to 600bhp. The Meteor Mk 4B was fitted to Centurions Mk 3 to Mk 10: various modifications were incorporated, including a modified fan drive, full-flow oil filter and increased capacity oil pumps, giving a power output of 620–630bhp. The Meteor Mk 4B/1 had larger main jets and chokes to the carburettor, roller rockers and a modified magneto advance, bringing the power output up to 650bhp. Finally, the Meteor Mk 4C had shaft-driven cooling fans. There was also a sixth engine variant, the Meteor Mk 4B/1/H, consisting of a Mk 4B/1 that

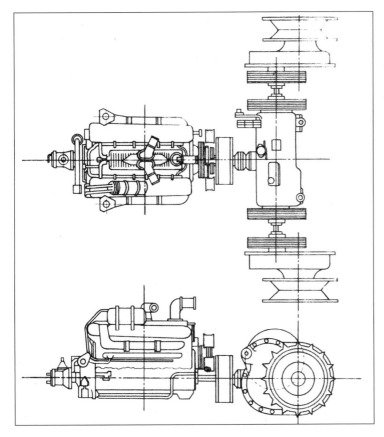

FVRDE diagram showing the arrangement of the main engine and transmission (drawing FV231921, sheet 5). (*Warehouse Collection*)

had been modified to provide drive for the hydraulic system operating the 'dozer blade of the AVRE.

All Meteor engines from the Mk 3 onwards had used two-piece cylinder blocks consisting of a cylinder skirt and a separate cylinder head, and all had four valves per cylinder, operated by twin overhead camshafts on each cylinder bank. The pistons ran in dry-fitted steel liner inserts. Lubrication was achieved by means of a gear-type pressure pump drawing oil from a pair of tanks, feeding it to the crankshaft and valve gear via a felt-element filter unit and finned oil cooler, before returning it to a shallow sump and thence back to the reservoir tanks via a pair of scavenger pumps. During 1954 there had been experiments using a hand-primed oil pump before starting the engine with a view to extending engine life. Two Meteor Mk 4B engines were used in the trial, one fitted to Centurion 11BA11, the other to 01ZR53. Both engines were fitted with a pump that allowed the lubricating system to be primed to 10lbf/in² before starting. The process was described as 'tedious' even before the trials were initiated, but in any case the trial was inconclusive since the engine in 01ZR53 caught fire in December 1954, while that fitted into 11BA11 showed greater than normal levels of wear because the driver had failed to change either the engine oil or the filters. The driver was disciplined and the trial was abandoned!

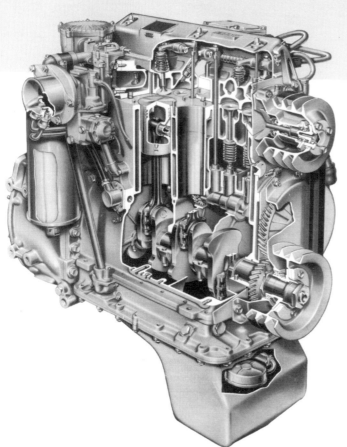

No matter how good the filtration system, the amount of dust kicked up by a tank on the move causes premature wear of the engine. Engine life in service was reported to be only around 3,000 miles before a base overhaul was required. The photograph shows a Swiss Army Pz 57/60 armed with the 150mm L7 gun. (*Simon Thomson*)

Cut-away view of the Rolls-Royce B40 four-cylinder petrol engine used to power the hydraulics of the FV4002 bridgelayer. A similar eight-cylinder (B80) unit was used to power the winch of the FV4006 armoured recovery vehicle (ARV). (*Rolls-Royce Limited*)

The engine was started by a 24V starter motor mounted directly on the wheelcase, driving the crankshaft via a spring drive.

Engine life in service was reported to be around 3,000 miles before a base overhaul was required, but in early 1961 the BAOR Operational Research Section published a report looking at the 'optimum life to base overhaul' of the Centurion in which 'consideration was given to the preventive exchange of assemblies at various vehicle mileages, and the effect this would have upon the overall efficiency of a tank force'. The report showed that nearly three times as many failures occurred in the second half of a tank's life and it was suggested that preventive exchange of the engine (as well as the gearbox, clutch and final drive) at 1,500, 2,000 or 2,500 miles would reduce the incidence of failure. However, it was also pointed out that, although many random failures could be avoided by such pre-planned replacements, the cost would be 'considerable' and there would be a loss of 'potential life' of assemblies that were rebuilt needlessly. The report proposed that the exchange of assemblies at 1,500 miles would be costly and impractical, but it was suggested that a reduction to 2,500 miles in the distance covered before a base overhaul was required would be a feasible and worthwhile step towards improving operational reliability.

The minimum fuel rating requirement was 80-octane – the so-called 'pool petrol' of the period – with the fuel pumped from a pair of internal tanks, one located either side of the engine, by twin David-type mechanical pumps. Combustion air was drawn through the louvres on the engine deck. A Ki-Gass priming pump was supplied to aid cold-weather starting. The fuel was fed through a bowl-type filter to a pair of Zenith twin-choke up-draught carburettors mounted on coolant-jacketed manifolds placed between the cylinder banks. Back in 1946 Centurion prototype number nine had been experimentally fitted with a mechanical fuel-injection system but it was never considered to be satisfactory, and the Zenith carburettors remained in place through the service life of the vehicle. The original fuel capacity was 121 gallons, but a third, armoured, fuel tank, giving an additional 109 gallons capacity, was installed when the Centurion Mk 7 was introduced in 1963. Refuelling the tanks from jerrycans and funnels while away from a proper base was a messy business that usually resulted in fuel spilling onto the top of the hull and onto the ground around the tank, often creating a real fire hazard.

Sparking was achieved via twin Simms magneto distributors, described as A and B, connected to radio-screened sparkplugs. To discourage over-revving, the distributor rotor arms were fitted with spring-loaded cut-outs that shorted out the sparks should the engine reach 2,250rpm. There were two plugs to each cylinder, those on the inlet sides of the blocks connected to magneto A and those on the exhaust side connected to magneto B.

Although the Meteor engine included a 1.5kW generator, belt-driven by the main engine, there was also an auxiliary engine designed to run a separate generator set and to drive the cooling fans; both the main and auxiliary engines were water-cooled on a common circuit, and the auxiliary engine was coupled to a pair of large-diameter fans drawing cooling air through the engine deck louvres and forcing it out through two gilled-tube radiators above the gearbox. The auxiliary engine was a four-cylinder Morris USHNM Mk 2 or Mk 2/1 side-valve unit of 918cc similar to that used in early MM Series Morris Minors, although the cylinder head and flywheel were quite different. Located in the engine compartment, the engine was coupled by splined shaft to a CAV DM12A/2 or DM12A/3 3kW generator producing a controlled 100A at 27V.

Access to the engine compartment was gained via five hinged and interlocking louvred covers that protruded above the general level of the top of the hull.

Power was conveyed from the main engine through a mechanically operated Borg & Beck 16in triple-plate dry clutch, mounted on a sub-frame, to a Merritt-Brown Z51R combined gearbox and steering unit, transversely mounted in a separate compartment – the 'R' suffix to the code indicated that the gearbox included an additional high-speed reverse gear. Designed during the 1930s by Dr H.E. Merritt, Director of Tank Design at Woolwich Arsenal, and patented and manufactured by the David Brown Company, the Merritt-Brown transmission was first trialled in the A20 infantry tank before being put into production for the Churchill. The system incorporated three differentials, and allowed the tank to be steered by changing the relative speeds of the two tracks while remaining under power. The gearbox offered five forward speeds and two reverse, as well as a differential lock. Each of the gears gave a different turning radius, ranging from 15ft in first gear to 130ft in fifth, and the steering remained operational even when the gearbox was in neutral, enabling the Centurion to rotate around its own axis, making a so-called neutral turn. Gear changes were made manually and required the driver to double declutch when changing down, meaning that it was all too easy to miss a gear change altogether, and this wasn't helped by the fact that the clutch was extremely heavy to operate, requiring considerable foot pressure (some 60lb) on the pedal. In practice, drivers discovered that rapid gear changes could be made by taking advantage of the clutch brake. Described as a 'stick change', this was done by pulling slightly on one of the steering levers while operating the gear-change lever; although it resulted in slightly erratic progress, with the tank twitching slightly at each gear change, there was no appreciable loss of momentum.

The drums for the steering brakes were located to either side of the gearbox on the shafts that conveyed power to the final-drive assemblies. The latter were of the double-reduction spur gear type and, together with the drive sprockets, were mounted at the rear. The final drive ratio was reduced from 6.94:1 to 7.47:1 from the Mk 2 on, in order to compensate for the additional weight; this had the effect of lowering the maximum road speed.

During the Second World War most British-designed tanks had been fitted with the Christie suspension system, in which the tank was able to achieve a higher speed across country because of the relatively long travel allowed by the combination of swinging arms and horizontal springs. It was felt that this would not be suitable for the Centurion due to the weight of the vehicle, and the Centurion was supported on six modified Horstman suspension units bolted directly to the hull and described as AEC-Rackham units. There were three bogie units on each side, each of which carried two pairs of 31.6in diameter rubber-tyred road wheels supported on radial arms and sprung by means of three horizontal concentric coil springs, guided by a central rod and tube. Hydraulic telescopic shock absorbers were incorporated in the front and rear suspension units. During 1953/54 experiments were carried out

Front and rear views of one of the Centurion AEC-Rackham-modified Horstman suspension units. The road wheels were supported on radial arms and sprung by means of three horizontal concentric coil springs, guided by a central rod and tube; hydraulic telescopic shock absorbers were incorporated in the front and rear suspension units as seen here. (*Warehouse Collection*)

in redesigning the shape of the sidewalls of the road wheel tyres with a view to reducing the occurrence of track-shedding and, following side-by-side trials of the original and the redesigned tyre, the new design, which was manufactured by Dunlop, was adopted. The modification increased the width of the tyre at the crown by 0.4in.

Removable armoured side plates – sometimes described as 'bazooka plates' – were fitted over the suspension units and the upper runs of the tracks, with three separate plates to either side.

Although there had been some early experiments with hydraulic disc brakes, these were found to offer little real advantage and the main brakes consisted of nothing more sophisticated than Girling-Bendix twin leading-shoe drum brakes, mechanically operated via the driver's floor pedal. The brake drums were bolted to the final-drive input shafts, additional to, and outboard of, the steering brakes. The lack of any servo assistance meant that the brakes generally required a pretty hefty shove and, if the driver missed a gear change on an upgrade, were sometimes not sufficiently powerful to prevent the tank running out of control.

The 24in-wide pin-jointed tracks were of cast manganese steel, consisting of 108 shoes on either side, with a pitch of 5.5in. Cast spuds on the outer face allowed the track shoes to dig into the ground for traction, while a cast horn on the inner face guided the track across the twin road wheels. Some Centurions operated by the Canadian and the Royal Netherlands armies were retro-fitted with rubber track pads, the latter in 1964, and from 1973 Danish Army Centurions were fitted with a track produced locally by Varde Stålfabrik A/S. Track adjustment was achieved by moving the front idler wheel on its eccentric axle, and the specified track tension required a 'sag' of 0.5 to 1in in the upper run ... in Korea it was found that at this setting the tank was more susceptible to shedding tracks, particularly on side slopes, and one group of users suggested that the track tension could be increased considerably if the tank was nudged, nose first, up against a bank before the idler wheel was adjusted, a practice that REME described as downright abuse! Sprocket life – providing the tracks were not over-tightened – was said to be in the order of 600–650 miles before the sprockets needed to be turned through 180 degrees; as the track stretched, links were removed, with the whole track considered to be ready for replacement once the number of links was reduced to 101.

A double track-return roller was carried on the top face of the first and third suspension units, with two more rollers attached to the hull either side of the centre suspension unit. Single rollers were also fitted immediately behind the drive sprocket at the rear and the idler wheel at the front, supporting only the inner sides of the track.

Firepower was always one of the great strengths of the Centurion and during the production life of the vehicle three different main guns were used, each representing an increase in penetrative power in response to perceived threats from Soviet armour.

The first of these was the Ordnance Quick Firing (OQF, or just QF) 17-pounder, a 76.2mm calibre rifled gun with a barrel-to-calibre ratio of L/55: this was fitted to both the Mk 1 and the Mk 2 variants. Developed during the Second World War the 17-pounder had already been used to equip the hybrid Anglo-American Sherman Firefly and, in a modified form, was also fitted into the British Comet tank. The term 'Quick Firing' meant that the propellant charge was loaded into a metal (in this case brass) case that sealed the breech to prevent escape of the expanding propellant gas. The gun was loaded from the side via a sliding breech, using fixed one-piece ammunition, and was capable of firing both high-explosive (HE and HEAT) and armour-piercing discarding sabot (APDS) rounds. Muzzle velocity with APDS rounds was 3,950ft/s, allowing the gun to penetrate up to 208mm of rolled homogenous armour (RHA), and the 17-pounder effectively outperformed all other Allied anti-tank guns of the period, and was one of the few Allied tank guns capable of penetrating all but the thickest areas of armour on the German Tiger and Panther tanks.

By 1947 the 17-pounder gun had been superseded by the much-improved Ordnance Quick Firing (OQF) 20-pounder 'tank gun, Mk 1', an 83.4mm L/66.7 rifled gun that made more than a nod in the direction of the German 88mm KwK 43 tank gun, the most-feared tank gun of the Second World War. Originally conceived as a 21-pounder (120mm), the new gun had been in development since 1945, and was capable of firing armour-piercing discarding sabot (APDS), armour-piercing cap ballistic capped (APCBC), high explosive (HE and HEAT) and solid shot rounds. Recoil was handled by a pair of buffer cylinders positioned to either side of the gun cradle, and, like the 17-pounder, the gun had a sliding breech and was loaded from the side using fixed ammunition. A modification was required soon after the gun's first use when it was found that the foil used in the manufacture of high-explosive rounds could cause the breech to jam. Firing was electrically controlled, and spent rounds were discharged automatically into a turret bin before subsequently being ejected manually from the tank through the port in the loader's side of the turret. When firing APDS rounds, the muzzle velocity was 4,810ft/s, allowing the gun to penetrate up to 300mm of RHA, a near 50 per cent improvement over the 17-pounder. The new gun was fitted into the Centurion Mk 3 and in some of the variants of the Mks 5, 7 and 8; in its so-called 'B-barrel' configuration, a fume extractor was fitted halfway down the barrel.

The last of the main guns fitted to the Centurion was the Royal Ordnance L7, an accurate and hard-hitting 105mm rifled gun that had been developed at the Royal Armament Research and Development Establishment (RARDE) at Fort Halstead in direct response to the Soviet T-54A medium tank, which carried a 100mm gun.

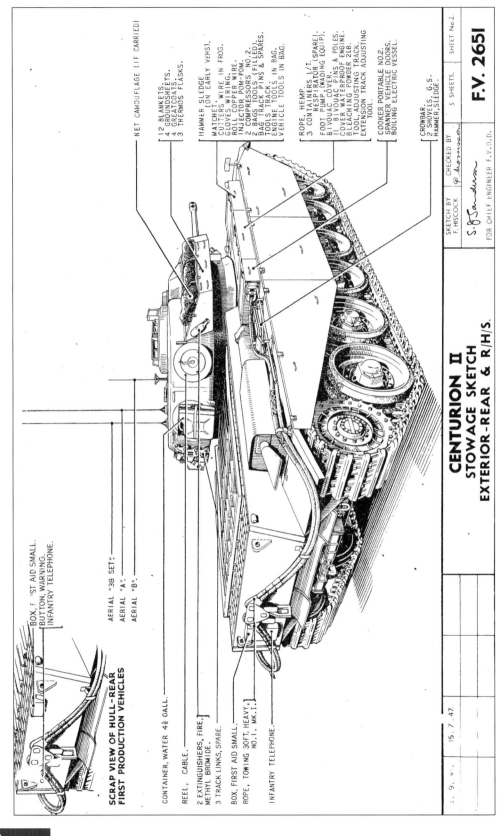

SCRAP VIEW OF HULL-REAR
FIRST PRODUCTION VEHICLES

BOX, F 'ST AID SMALL.
BUTTON, WARNING.
INFANTRY TELEPHONE.

AERIAL "38 SET".
AERIAL "A".
AERIAL "B".

NET CAMOUFLAGE (IF CARRIED)

12 BLANKETS.
4 GROUNDSHEETS.
4 GREATCOATS.
3 THERMOS FLASKS.

HAMMER SLEDGE
(ON EARLY VEHS).
MATCHET.
CUTTERS WIRE IN FROG.
GLOVES COPPER WIRE.
ROLL.
INJECTOR POM-POM.
2 COMPRESSORS NO.2.
2 BAGS TOOLS (FILLED).
BAG TRACK PINS & SPARES.
TOOLS TRACK.
ENGINE TOOLS IN BAG.
VEHICLE TOOLS IN BAG.

ROPE, HEMP.
3 CONTAINERS L/T.
RESPIRATOR (SPARE).
FOOT PUMP.(WADING EQUIP).
BIVOUAC COVER.
II BIVOUAC PINS & POLES.
COVER WATERPROOF ENGINE.
BLEACHING POWDER 2LB.
TOOL, ADJUSTING TRACK.
EXTENSION TRACK ADJUSTING
TOOL.

COOKER PORTABLE NO.2.
SPANNER VEHICLE DOORS.
BOILING ELECTRIC VESSEL.

CROWBAR.
2 SHOVELS, G.S.
HAMMER, SLEDGE.

CONTAINER, WATER 4½ GALL.

REEL, CABLE.

2 EXTINGUISHERS, FIRE.
METHYL BROMIDE.

3 TRACK LINKS, SPARE.

BOX, FIRST AID SMALL.

ROPE, TOWING 30FT. HEAVY,
NO.1, MK.I.

INFANTRY TELEPHONE.

5. 9. 4.	15. 7. 47.	—	—	CENTURION II	SKETCH BY F. HISCOCK	CHECKED BY P. Normaou	SHEET No.2
				STOWAGE SKETCH EXTERIOR-REAR & R/H/S.	S. J Jardina		5 SHEETS.	F.V. 2651
					FOR CHIEF ENGINEER F.V.D.D.			

British Army stowage diagram showing items stowed on the exterior rear and right-hand side of the Centurion Mk 2. *(Warehouse Collection)*

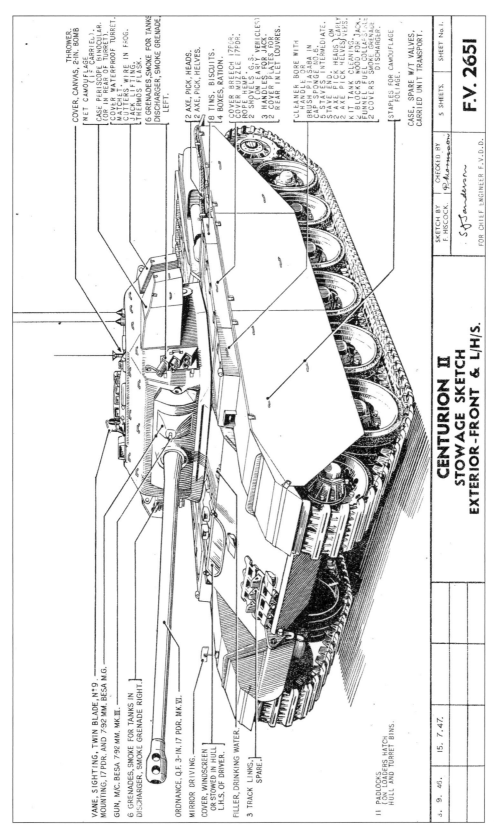

VANE, SIGHTING, TWIN BLADE, Nº 9.
MOUNTING, 17 PDR. AND 7·92 MM. BESA M.G.

GUN, M/C. BESA 7·92 MM. MK. III.

6 GRENADES, SMOKE FOR TANKS IN
DISCHARGER, SMOKE GRENADE RIGHT.

ORDNANCE, Q.F. 3-IN. 17 PDR. MK. VI.

MIRROR DRIVING.

COVER, WINDSCREEN
OR STOWED IN HULL
L.H.S. OF DRIVER.

FILLER, DRINKING WATER.

3 TRACK LINKS,
SPARE.

11 PADLOCKS
(ON LOADERS HATCH
HULL AND TURRET BINS.)

COVER, CANVAS, 2-IN. BOMB
THROWER.
NET CAMOUFLAGE.
(IF CARRIED.)
CASE PERISCOPE BINOCULAR.
(FOR IN REAR OF TURRET.)
COVER WATERPROOF TURRET.
MATCHET.
CUTTERS, WIRE IN FROG.
JACK LIFTING.
THERMOS FLASK.

6 GRENADES, SMOKE FOR TANKS
DISCHARGER, SMOKE GRENADE,
LEFT.

2 AXE, PICK, HEADS.
2 AXE, PICK, HELVES.

8 TINS, BISCUITS.
4 BOXES, RATION.

COVER BREECH 17PDR.
COVER MUZZLE 17PDR.
ROPE HEMP.
2 SHOVELS, G.S.
(ON EARLY VEHICLES)
3 HANDLES FOR JACK.
2 COVER PLATES FOR
REAR INLET LOUVRES.

CLEANER BORE WITH
HANDLE OR
BRUSH PIASABA IN
CAP SPONGE NO. 6.
5 STAVES INTERMEDIATE.
STAVE END.
2 AXE PICK HEADS EARLY
2 AXE PICK HELVES VEHS.
KIT TANK CLEANING.
6 BLOCKS WOOD FOR JACK.
FUNNEL FUEL COLLAPSIBLE
2 COVERS SMOKE GRENADE
DISCHARGER.

STAPLES FOR CAMOUFLAGE
FOLIAGE.

CASE, SPARE W/T VALVES,
CARRIED UNIT TRANSPORT.

CENTURION II
STOWAGE SKETCH
EXTERIOR-FRONT & L/H/S.

3. 9. 46.	15. 7. 47.	SKETCH BY F. HISCOCK.	CHECKED BY P. Monnason	5 SHEETS.	SHEET No. I.
		S J Sanderson			**F.V. 2651**
		FOR CHIEF ENGINEER F.V.D.D.			

British Army stowage diagram showing items stowed on the exterior front and left-hand side of the Centurion Mk 2. (Warehouse Collection)

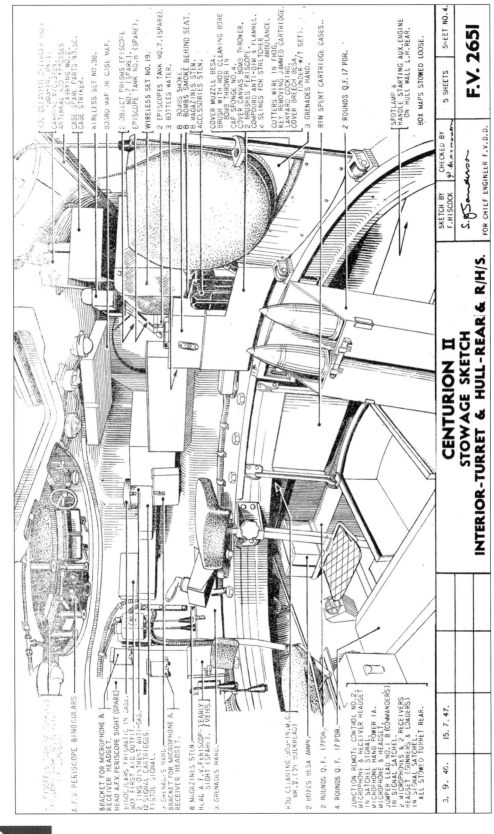

CENTURION II
STOWAGE SKETCH
INTERIOR-TURRET & HULL-REAR & R/H/S.

SKETCH BY F.HISCOCK	CHECKED BY P.M.Iverson
S.F.Saunders	FOR CHIEF ENGINEER F.V.D.D.

	5 SHEETS	SHEET NO.4.
		F.V. 2651

Interior stowage diagram showing items carried in the rear of the turret and the fighting compartment of the Centurion Mk 2. (Warehouse Collection)

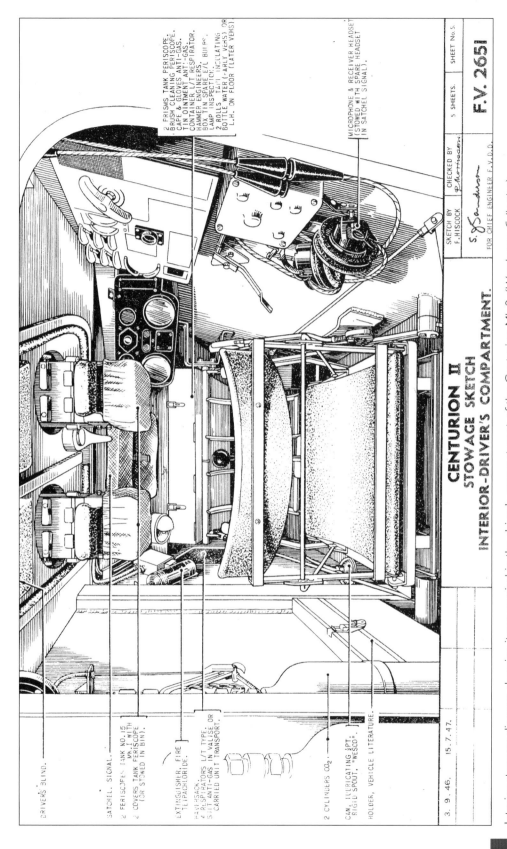

DRIVER'S BLIND.

SATCHEL, SIGNAL.

2 PERISCOPES TANK NO. 15
2 COVERS TANK PERISCOPE 1½" WITH
(OR STOWED IN BIN).

EXTINGUISHER, FIRE
TETRACHLORIDE.

HAVERSACK.
4 RESPIRATORS L/T TYPE.
SUIT ANTI-GAS IN VALISE OR
CARRIED UNIT TRANSPORT.

2 CYLINDERS CO₂.

CAN, LUBRICATING ½PT.
RIGID SPOUT, "WESCO".

HOLDER, VEHICLE LITERATURE.

2 PRISMS TANK PERISCOPE.
BRUSH CLEANING PERISCOPE.
CAPE & GLOVES ANTI-GAS.
CONTAINER ANTI-GAS.
OINTMENT RESPIRATOR.
HAMMER ENGINEERS.
BOX TIN SPARE E/L BULBS.
LAMP INSPECTION.
2 ROLLS TAIL INCLUDING
BOTTLE WATER (EARLY VERS.) OR
L.H. ON FLOOR (LATER VEHS)

MICROPHONE & RECEIVER HEADSET
(STOWED WITH SPARE HEADSET
IN SATCHEL SIGNAL).

3. 9. 46.   15. 7. 47.

		SHEET No. 5.
SKETCH BY F. HISCOCK	CHECKED BY P. Mortiboer	5 SHEETS.
S. Sandiver		
FOR CHIEF ENGINEER F.V.D.D.		**F.V. 2651**

CENTURION II
STOWAGE SKETCH
INTERIOR—DRIVER'S COMPARTMENT.

Interior stowage diagram showing items carried in the driver's compartment of the Centurion Mk 2. (Warehouse Collection)

79

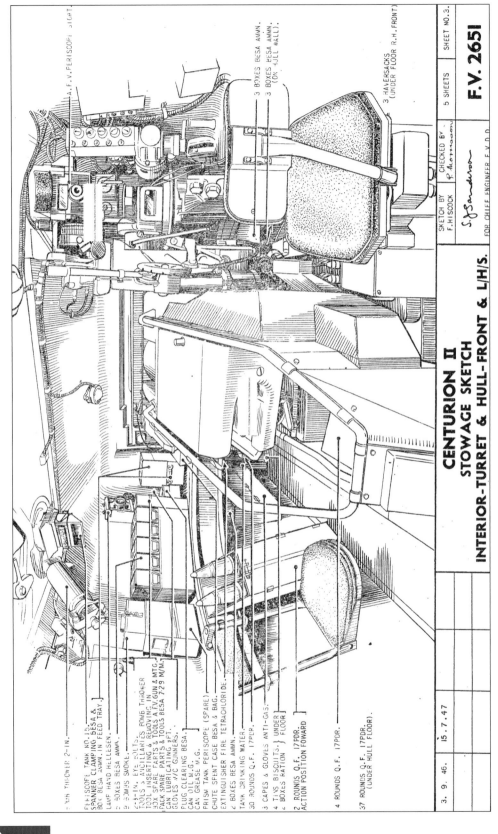

CENTURION II
STOWAGE SKETCH
INTERIOR–TURRET & HULL–FRONT & L/H/S.

F.V. 2651

| SKETCH BY F.HISCOCK | CHECKED BY P. Morrison | 5 SHEETS | SHEET NO.3. |

S.J.Sanders
FOR CHIEF ENGINEER F.V.D.D.

| 3. 9. 46. | 15 . 7 . 47 | | | |

Interior stowage diagram showing items carried in the front of the turret and the fighting compartment of the Centurion Mk 2. (*Warehouse Collection*)

Developed by increasing the bore of the 20-pounder, the L7 was a high-velocity 105mm L/52 weapon with an effective rate of fire of eight to ten rounds per minute and was designed to fit directly into the same mounting, allowing existing Centurions to be up-gunned with the minimum modification ... at a price of around £2,100 per tank. By the simple expedient of swapping the gun barrels, the L7 also allowed stocks of 20-pounder ammunition to be used up during live firing practice. The L7 made its first appearance in the Centurion in 1959, when it was fitted to the Mk 5/2, and went on to be used in some variants of the Mks 7 and 8, and in all variants of the Mks 9 to 13, becoming the de facto standard NATO tank gun during the 1960s. Like the 'B-barrelled' 20-pounder weapons, the L7 was fitted with a fume extractor mounted halfway along the barrel, this time in an eccentric position that allowed greater gun depression over the engine covers at the rear; at the same time that the coaxial ranging machine gun was fitted (see below), the L7 was fitted with a thermal sleeve designed to reduce barrel droop during prolonged firing and thus increasing long-range accuracy.

As with the 20-pounder, the gun was triggered electrically, and suitable rounds included armour-piercing discarding sabot (APDS and APDS-T), armour-piercing fin-stabilised discarding sabot (APFSDS and APFSDS-T), anti-personnel (APERS and APERS-T), high explosive (HE and HEAT), and high-explosive squash head (HESH); the '-T' suffix indicates a tracer round. Using APFSDS rounds the muzzle velocity was 4,868ft/s, giving a maximum penetration of 340mm of RHA. When firing APDS rounds the effective range was 1,800 yards, and with HESH rounds the range was extended to 3,000–4,000 yards.

Regardless of 'mark', all Centurion gun tanks had the ammunition for the main gun stowed in such a way as to reduce the risk of fire if the tank was hit, a problem that had been all too common during the Second World War. Something like half of the available ammunition was stowed in the front left-hand side of the hull, with the remainder under the turret floor, in bins on the turret ring, and under the gun. See Table 1 (p. 120) for details of ammunition stowage.

Finally, mention must also be made of the Centurion AVRE (assault vehicle, Royal Engineers), which carried a Royal Ordnance BL 165mm L9A1 demolition charge projector capable of firing a 64lb high-explosive squash head (HESH) round. Although the gun's muzzle velocity was low, the ammunition, which contained 40lb of C4 plastic explosive, was sufficiently powerful to demolish targets such as pillboxes, roadblocks, bunkers and buildings, and the gun was accurate to about 2,500 yards.

Vision and sighting equipment consisted of a raised all-round vision (ARV) rotating cupola for the commander that allowed him to lay the gun on a target. Three different patterns of cupola were fitted, according to the mark of the tank, with that fitted to the Centurion Mk 8 contra-rotating. The cupola was fitted with seven, eight or nine

A Leyland-built Centurion Mk 3, subsequently converted to Mk 10 configuration, being put through its paces at the Tank Museum's Tankfest event. (*Simon Thomson*)

episcopes, according to type, and there was also a periscopic sight with ballistic pattern graticules injected from the gunner's sight, and a 10x magnification periscopic binocular. A collapsible vane sight was fitted to the turret roof directly in front of the vision cupola and this was used by the commander to lay the gun for both indirect and semi-indirect shooting, with the gunner subsequently using the traverse indicator and the clinometer attached to the sight mechanism. For direct shooting, the gunner firstly used the turret traverse and then set the estimated range on a drum that was engraved to cover 3,000–8,000 yards. There was also a 6x magnification periscopic sight for the gunner with ballistic pattern graticules, and a periscope was provided for the loader alongside the roof-mounted escape hatch.

The gun-control equipment offered four modes of operation – manual elevation and traverse, non-stabilised powered traverse, fully stabilised powered traverse and elevation, and emergency powered traverse, the latter with just a single speed. Described as 'fighting vehicle gun control equipment (FVGCE) number 1, Mk 4', the powered stabilising system equipment, manufactured by Metropolitan-Vickers (Metrovick), was electro-mechanical in operation, allowing accurate firing on the move by using a pair of gyroscopes to measure deviations in the movement of the tank in relation to the original gun setting, with necessary corrections transmitted via an amplifier to servomotors coupled to the gun mount and turret traverse. Although complex, the system proved to be very reliable, particularly when compared to hydraulic systems of the period. However, it was not totally automatic, and still required the gunner to make fine corrections to both azimuth and elevation.

As regards secondary weapons, the Polsten 20mm gun that had been fitted to some of the prototypes was generally considered to be unwieldy and too large for use against infantry and was soon abandoned as unsatisfactory, with very few examples of the Mk 1 A41* thus equipped. The standard secondary weapons throughout the production life of the Centurion were the Besa machine gun, effectively a British version of the Czech ZB-53 air-cooled belt-fed weapon with a calibre of 7.92mm, and the 0.30in Browning. For those variants where just one secondary gun was fitted, it was carried in a coaxial ball mount to the left of the main gun (when viewed from inside the turret), while the second weapon, if fitted, was carried in an anti-aircraft mount. From Mk 10/2 onwards, most variants were also fitted with an 0.50in Browning ranging machine gun designed to fire sets of three tracer rounds to allow the commander to accurately estimate the range of the target without wasting the more expensive rounds of the main gun. The gun was also coaxially mounted, between the main gun and the 0.30in machine gun. See Table 2 (p. 121).

Chapter Six

Improving the Breed

While the Mk 13 may have marked the end of the 'official' development of the Centurion, the numbers of Centurions in use around the world were such that development didn't halt when the last example was constructed and, even without new sales, the Centurion remained big business with money to be made from selling upgrades. A number of nations operating Centurions put the tanks through an improvement programme using either their own or third-party equipment.

Most notable among the improved Centurions were the tanks produced by the Israeli Ordnance Corps Workshops at Tel a Shumer. The Israeli Defence Force (IDF) had taken delivery of a modest number of Centurions in 1959, naming them Sho't, simply meaning Whip in Hebrew. The vehicles were said to have been unpopular with crews who were more used to the simpler M4 Sherman tank, while there was also some dissatisfaction with the low power-to-weight ratio and resulting high fuel consumption, and the difficulties that were encountered with the cooling and filtration systems when the tank was operated in desert terrain. The decision was ultimately taken to replace the Meteor engine with a more economical diesel power unit, and, following trials with three different types of engine, at the beginning of the 1970s the first of the Israeli-upgraded Centurions started to enter service. The engine chosen was the Teledyne Continental AVDS 1790-2AC, an American-built air-cooled diesel engine that was at least partly chosen for its commonality with the power unit fitted to the Israeli upgraded M48A2 tanks; there was also a new Allison CD-850-6A automatic transmission. The rear of the hull was raised to accommodate the new engine, which nevertheless had to be installed at an angle. Other improvements included increased fuel capacity, a more effective braking system, increased and improved ammunition stowage, a new electrical system and improvements to the fire-fighting system that prevented the extinguishing agent from being prematurely dispersed by the engine cooling fans. Although officially designated simply as 'upgraded Centurion', the vehicle was also described as Sho't Kal, the 'Kal' designation being a Hebrew abbreviation of the engine manufacturer's name, Continental. (See *colour plate 7, a and b*.) By 1974 all Israeli Centurions had

(*Above*) Externally clad in appliqué reactive armour panels, this Centurion Mk 10 (FV4017) of the Swedish Army has been up-gunned with the 105mm L7 weapon and fitted with the same power pack and transmission as the Israeli Sho't Kal, making the designation Strv 104. Note the locally sourced lighting equipment. (*Simon Thomson*)

(*Opposite, top*) The Nagmachon (the name simply means 'underbelly-protected') is a heavily armoured personnel carrier (APC), one of a number of armoured engineers' vehicles deployed by the Israeli Defence Force (IDF). The raised superstructure is sometimes referred to as a 'doghouse'. (*Natan Flayer*)

(*Opposite, below*) Photographed at the Eurosatory 2004 defence show in Versailles, this photograph shows the Centurion-based Puma heavy armoured engineering vehicle equipped with the Israeli Carpet minefield clearance system, consisting of twenty rockets with a fuel-air explosive warhead that spreads a cloud of fuel fumes which, when detonated, will destroy most types of mine. The rockets can be fired singly or all together. (*Simon Thomson*)

been upgraded to Sho't Kal configuration, as well as being improved in various other ways, including the use of Blazer explosive reactive armour (ERA) designed to disrupt high-explosive armour-piercing (HEAT) rounds.

From about 1978/79 the Centurion gun tank started to be replaced by the indigenous Merkava in IDF armoured units, and withdrawn Sho't and Sho't Kal Centurions were converted to heavy armoured personnel carriers (APC) for use by combat engineers. Designated Puma, NagmaSho't, Nakpadon and Nagmachon, the vehicles were heavily protected by composite and explosive reactive armour.

A Centurion upgrade kit was also developed by the Israeli company NIMDA, based in Tel Aviv, specifically for the Centurion Mk 5. The kit included a Detroit Diesel Allison (DDA) 12V-71T turbocharged diesel engine and an Allison torque-converter transmission. The standard OQF 20-pounder gun was replaced by an American 105mm M68 gun, manufactured by Watervliet Arsenal; effectively, this was the British 105mm L7 gun with a new drop-block breech. And, still in Israel, an improved commander's cupola was produced for the Centurion by Menachem Urman and Company of Yahud. Developed for the Israeli Defence Force, the cupola provided full 360-degree vision combined with maximum head protection.

In the USA in the late 1980s and early 1990s NAPCO International of Minnesota offered a retrofit diesel power pack and automatic transmission system for older tanks, with kits available for the Centurion as well as the M4 Sherman and the M47 and M48 Patton tanks. As with the NIMDA kit, the engine was the Detroit Diesel Allison (DDA) 12V-71QTA, fitted with four turbochargers and an Airscrew Howden cooling system; the transmission was the Allison CD-850-6.

A similar upgrade package, providing a 50 per cent improvement in fuel consumption and an increased top speed, was devised by Vickers Defence Systems (previously Vickers-Armstrongs) and Marconi Radar Systems. The package included a turbocharged diesel engine, either a Rolls-Royce CV12 TCE or a Detroit Diesel Allison 12V-71T, coupled either to the original Merritt-Brown Z51R gearbox, or to a Self Changing Gears TN12 hot-shift semi-automatic transmission. There was an optional NBC (nuclear-biological-chemical) protection kit, and improvements were also made to the commander's cupola, and to the fire-control and stabilisation systems. The package could also embrace replacement of the original gun with a 105mm L7 weapon. Two prototypes were constructed for Centurion Mk 5s in service with the Swiss Army, and one for Sweden, but no further orders were forthcoming. However, in 1981 the Swedish government awarded a contract to Bofors for upgrading the Centurions Strv 101 and 102 (see *colour plate 5*) by fitting the Lyran illuminating rocket flare system, as well as making improvements to the fire-control and sighting equipment. Some eighty-four examples of the Strv 102 were also subsequently upgraded to Strv 104 configuration using the same power

The final version of the Swiss Army Centurion, fitted with the 105mm L7 gun, was the Panzer 57/60. (*Warehouse Collection*)

pack and transmission as the Israeli Sho't Kal; some were also fitted with reactive armour.

Jordanian Centurions were also re-engined using the Teledyne AVDS 1790-2AC air-cooled diesel engine, as well as being fitted with a license-built Hughes fire-control system with a laser rangefinder. In Switzerland the Saurer company (Aktiengesellschaft Adolph Saurer) also developed an engine upgrade for the Centurion that replaced the original Meteor engine with an MTU MB 837EA 500 turbocharged diesel engine producing 750bhp, together with a fully automatic gearbox. The package was said to require the minimum of changes to the engine compartment.

Mention must also be made of the South African Skokiaan, Semel and Olifant tanks, all of which were based on a modified Centurion. Dating from 1972, the Skokiaan (the name describes an illicit alcoholic beverage favoured in the South African townships) was the first of these upgrades, with eight tanks fitted with a new V12 fuel-injected engine, together with a new three-speed (3F2R) automatic transmission. Two years later the Skokiaan project was followed by the conversion of thirty-five tanks, including the eight tanks that had already been re-engined, plus other Centurions that had been bought from Indian scrap yards, into what was described as Semel (Tempest) configuration. Similar to the Israeli Sho't, the Semel tanks were fitted with a Teledyne Continental AVDS 1790-2AC air-cooled diesel engine and an Allison CD-850-6A automatic transmission.

Finally, the Olifant (Elephant) was developed by the South African Vickers OMC company, formerly Reumech OMC, to counter the possible deployment of the Soviet T-72 tank in Southern Angola. Work started in 1976 and the tank was produced in four variants, identified as Olifant Mks 1, 1A and 1B, and Olifant Mk 2. The first to be produced, the Olifant Mk 1, was fitted with improved suspension, a more efficient turret-traverse system and night-sights. Development of the Olifant Mk 1A started in 1983, with the first vehicles delivered two years later. It was up-armoured and fitted with a new, locally produced air-cooled V12 diesel power pack; the old 20-pounder, which had proved ineffective against Soviet T-55s, was replaced by the 105mm L7 gun; smoke dischargers were fitted on either side of the turret; rangefinder and night-vision equipment was also improved. The Mk 1B entered production in 1991, and was also up-armoured, particularly on the hull floor, and was fitted with a thermal sleeve on the gun barrel. Produced by the LIW Division of the South African company Denel Land Systems (Pty), the Olifant Mk 2 was basically an up-armoured Mk 1B in which the main gun had been replaced by a 120mm smooth-bore weapon, together with further improvements to the fire-control equipment. The Olifant Mk 2 remained in service with the South African 1st Tank Regiment until at least 2010.

Chapter Seven

Centurion Engineers' Tanks

Various types of armoured support vehicle were developed during the Second World War, and these contributed in no small part to the success of the Allied invasion of Europe in June 1944. The range of vehicles included 'dozer tanks, armoured recovery vehicles (ARVs), beach armoured recovery vehicles (BARVs), armoured engineers' vehicles (AVREs), bridgelayers (AVLBs) and armoured ramp carriers (ARKs), and various types of mine-clearance vehicle. Most were based on either the Sherman or Churchill hull and, despite their age, many remained in service with the British Army into the immediate post-war years. As these vehicles began to age, or became obsolete for other reasons, it was the Centurion chassis that was generally selected as the replacement.

The first to go was the old armoured recovery vehicle. When the Centurion started to enter service, it soon became apparent that a Sherman- or Churchill-based armoured recovery vehicle was not really going to be adequate to recover the near 50-ton weight of the new tank. Clearly a new recovery vehicle would be required. As an interim measure, a number of damaged Centurions were converted to tugs, or 'towers', in the British Army's Commonwealth Base Workshop at Kure, Japan. Some were used in Korea, both for recovery and as supply and ammunition carriers, where it was shown that they were capable of hauling sledges loaded with up to 2 tons of ammunition up mountain sides. Other tugs were constructed by modifying obsolete gun tanks at 27 Base Workshop in Britain and at 7 Armoured Workshop BAOR.

However, work had already started on designing a purpose-made armoured recovery vehicle using the Centurion hull. REME 13th Command Workshops (now 43rd) constructed a prototype at Aldershot during 1951 that was similar in concept to the old Churchill ARV Mk 2. There was a dummy turret and gun, and an 18-ton winch powered by the six-cylinder engine of a Bedford truck, but the crew compartment was cramped due to the need to provide a separate winch engine. Eleven examples had been constructed by the end of 1951, with a total of eight sent to Korea, and two to BAOR. Eventually, a total of 170 Centurions were converted to ARV Mk 1 configuration, with any problems that arose in service being ironed out as the conversion work continued.

(*Above*) A Centurion armoured recovery vehicle (ARV) Mk 2 (FV4006), showing the huge rear-mounted earth anchor that was deployed when using the winch. Differing in many respects from the REME-built Mk 1, the ARV Mk 2 was constructed by Vickers-Armstrongs using Mk 1, 2, 3 and 5 hulls. (*Warehouse Collection*)

(*Opposite, top*) Unloved and abandoned at REME's Bordon site, this ARV Mk 2 (FV4006) is almost certainly destined to end its days as a range target. (*Simon Thomson*)

(*Opposite, below*) This ARV Mk 2 (FV4006) is now in private hands – and registered for the road. The rubber track pads might suggest that this vehicle was originally used by either the Canadian or the Royal Netherlands Army, both of whom retro-fitted at least some of their Centurions with rubber track pads. (*Simon Thomson*)

Meanwhile, development work continued on the 'official' Mk 2 (FV4006) Centurion ARV. (*See colour plate 10.*) The first prototype for the Mk 2 (03ZR52) was constructed by Garner Motors of Acton in 1952, with trials continuing during 1953 and 1954. It was found that the removal of the turret reduced the weight of the machine to the point where its hill-climbing performance could be described as 'remarkable'. The Bedford petrol engine that had been used to drive the winch on the Mk 1 variant was abandoned in favour of an electric motor powered by a 400V 160Ah generator that was in turn driven by a Rolls-Royce B80 No. 1 Mk 2P or 5P eight-cylinder petrol engine, with a maximum power output of 165bhp at 3,750rpm from a displacement of 5,675cc. The generator was coupled to the winch via a roller chain. A huge spade anchor fitted at the rear was designed to be deployed using the main winch cable, and improvements were made to the layout of the winch, which was now rated at 30 tons for a direct pull, and to its roping arrangements, which now allowed pulls to the front or rear, or to either side by the use of pulley blocks. Considerable thought was given to the dissipation of heat produced by the winch and its power unit, which could result in overheating of the crew compartment, particularly when operating in high ambient temperatures. Although the problem was largely solved by means of fitting a large oil cooler together with a pair of fans forcing air through them, it was still necessary to limit the vehicle to five or six 'pulls' at full load. At the same time, trials were being carried out to determine the optimum construction of the steel wire rope used on the winch in order to obtain the best balance of life and wear.

A hull-mounted A-frame jib was also developed that could be attached to the front of the hull, allowing the ARV to remove and replace power packs in the field, and to provide a suspended tow to disabled vehicles.

Although a markedly better vehicle than the Mk 1, the hull remained cramped for the four-man crew, particularly the wireless operator, who had very little headroom when the hull was closed down, and it was occasionally necessary to remove the roof to gain access for maintenance of the winch. Production began in 1955 at ROF Woolwich and Vickers-Armstrongs, with the first example accepted into service in 1956; most of the 345 vehicles constructed were converted from obsolete gun tanks, and some were still in service as late as the end of the 1980s. In 1962 many of those in service with BAOR were modified to carry two spare 105mm gun barrels in place of the side stowage bins. This proved not to be entirely satisfactory because, although it reduced the need for accompanying trucks to carry the gun barrels, the ARV was not actually equipped to change the barrels and the attendance of a crane-equipped vehicle was still required.

A proposed Centurion ARV Mk 3 (FV4013) with a more spacious forward crew compartment, which would have placed the driver in with the rest of the crew, was never pursued beyond a design study.

During the amphibious landing stages of the D-Day assault there had been a second type of ARV designed exclusively for recovering drowned or disabled tanks and trucks on the landing beaches. Based on the M4 Sherman hull from which the turret and gun had been removed, and the sides raised to allow the vehicle to wade in up to 8ft of water, the beach armoured recovery vehicle (BARV or, occasionally, beach ARV), had shown itself to be enormously useful. The Sherman BARVs remained in service until the late 1950s but by this time they were proving unable to recover the heavier armoured vehicles that had started to enter service and, when the question of replacement arose in 1956/57, it seemed logical to use the hull of the Centurion. The Fording Trials Branch of REME used an obsolete Centurion 'tower' to produce a mild-steel mock-up along the lines of the Sherman BARV, operated by a four-man crew. The hull was extended forwards by about 5ft, with a large rope-cushioned pusher pad installed at the front,; this pad was subsequently replaced by a hardwood nosing block to reduce the danger of damaging landing craft. The prototype, which had been constructed from mild steel, was demonstrated at the Amphibious Trials and Training Unit (ATTU) at Instow in 1958/59 before being handed over to FVRDE for final development of the 'proper' armoured version. By the end of 1960 a batch of just twelve Centurion BARVs (FV4018) had been constructed at ROF Leeds using a mix of redundant Mk 1, Mk 2 and Mk 3 hulls. (*See colour plate 9.*) The overall height, with the armoured hull extension, was 140in, and the vehicle was capable of wading in a maximum of 114in of water, although at this depth the driver was effectively 'blind' and was obliged to rely on voice commands from the commander to direct the 40-ton machine. A set of lifting gear was also developed that could be attached to the hull of the BARV to enable it to lift out its own engine.

When the Army's amphibious capability was phased out in favour of the Royal Marines, the Centurion BARVs were similarly reassigned, and two are known to have been taken to the Falkland Islands in 1982 along with the British Task Force.

At around the same time that the Centurion hull was being pressed into service as a BARV, the finishing touches were also being put to a Centurion-based bridgelayer. Early development work on armoured bridgelayers – or armoured vehicle launched bridges (AVLB) as they tend to be described these days – had been conducted using the hulls of Valentine infantry tanks, but most of those constructed and deployed during the Second World War were carried on a modified Churchill hull. As regards its post-war replacement, it had originally been intended to use the chassis of the FV200 series universal tank but when this was abandoned work had started on developing a Centurion bridgelayer; as early as 1946 experiments had involved mounting a lattice steel framework onto the hull of Centurion prototype number three to test the manoeuvrability of what was

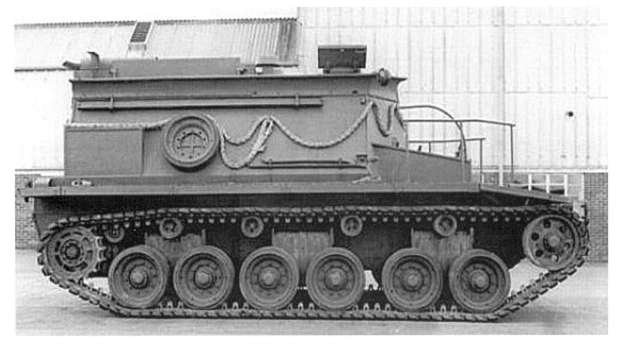

Designed to provide assistance to vehicles that had become waterlogged or bogged-down on landing beaches, the Centurion FV4018 beach armoured recovery vehicle (BARV) replaced the older Sherman-based machines. The raised sides to the hull allowed the vehicle to operate in water up to 114in deep. (*Warehouse Collection*)

Rear view of the FV4018 Centurion BARV. Interestingly, this example – one of twelve constructed by ROF Leeds on a mix of redundant Mk 1, Mk 2 and Mk 3 gun tank hulls – has rubber track pads. (*Warehouse Collection*)

Newly constructed Centurion BARVs (FV4018) awaiting issue; both of the examples seen here are based on the Mk 3 hull. The rope-cushioned pusher pad at the front was subsequently replaced by a hardwood nosing block. (*RAWHS*)

A REME BARV approaching an amphibious 'air portable general purpose' (APGP) Land Rover that has presumably been swamped. (*Warehouse Collection*)

inevitably a somewhat extended vehicle. In 1952 a mock-up bridge was mounted onto a Centurion Mk 1 production hull, and by 1956 a working prototype had been constructed by Hudswell Clark of Leeds using a Mk 2 hull. User trials of this, and a second, modified prototype, this time possibly based on a Mk 7 hull, were completed by September 1958. By this time it had already been decided that the production vehicles would be based on redundant Mk 3 or Mk 5 hulls that would be reworked to bring them up to Mk 7 standard.

The Class 80 'bridge, tank, number 6' consisted of four identical aluminium-alloy quarter sections topped with mesh trackways that were joined together in pairs to give a 52ft-long double-track bridge, capable of spanning 42ft. The bridge was carried in an inverted position along the length of the tank hull, and when the bridge was fully assembled the vehicle was almost 53ft long, so the bridge sections were generally carried on a small fleet of 3-ton trucks until required for deployment. In use, the four bridge sections were connected by two portal frames and a diagonal brace, and were attached to the launch arm at the nose of the vehicle. A lifting jib on the launch arm was used to assist in the assembly of the bridge sections, with brackets on the trackways used to attach the bridge to the arm. In-fill panels were carried on the sides of the hull, designed to be placed across the gap between the two longitudinal bridge components to allow smaller, wheeled vehicles to cross. During the launch operation, which took two minutes, the assembled bridge was simply lifted from its stowed position and rotated through 180 degrees vertically before being placed on the ground behind the tank, at which point it was disengaged from the mechanism and the tank would withdraw. Recovery was more or less the reverse of the launch process, taking four minutes.

The hydraulic equipment required to launch and recover the bridge was powered by a Rolls-Royce B40 No. 1 Mk 5P four-cylinder petrol engine, with a power output of 62bhp at 2,800rpm. The auxiliary engine was connected to a Towler Brothers hydraulic pump by a drive shaft taken directly from the flywheel, and was controlled electrically by solenoid-operated valves. Both the auxiliary engine and the hydraulic equipment were housed inside the fighting compartment.

The resulting vehicle weighed 49.6 tons with the bridge in place but nevertheless was capable of a top speed of 20mph on the road. The first pre-production Centurion bridgelayer was completed by ROF Leeds in early 1960 and, following acceptance trials, production proper started in 1961 and continued until 1963, with many going for export. The sheer size of the machine and the resultant lack of manoeuvrability meant that it was unable to keep up with the gun tanks in urban areas, not least in West Germany where most were deployed. Nevertheless, the vehicle remained in service until 1974, when it was replaced by the Chieftain bridgelayer that deployed a more versatile scissors bridge.

A Centurion bridgelayer (FV4002). A mock-up was constructed using a redundant Mk 1 hull, but production vehicles were built using Mk 3 (seen here) or Mk 5 hulls. The bridge was launched hydraulically simply by lifting it from the hull and rotating it through 180 degrees before disengaging the launch vehicle. With the four-part bridge in place, the overall length of the vehicle was more than 53ft. (*Warehouse Collection*)

A Centurion bridgelayer (FV4002) without the bridge; the launch arm and the bridge carrier can be seen at the front of the vehicle. In the transport position the bridge rests across the hull with the roadway facing downwards, and is supported by the beam at the rear. (*RAWHS*)

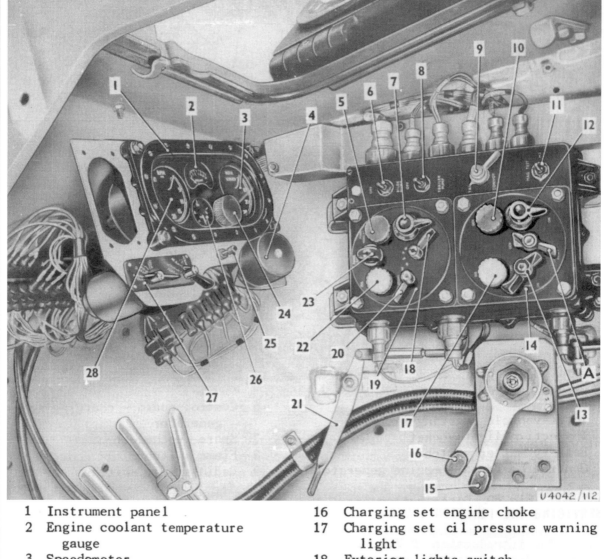

1	Instrument panel	16	Charging set engine choke
2	Engine coolant temperature gauge	17	Charging set oil pressure warning light
3	Speedometer	18	Exterior lights switch
4	Horn push and dipswitch	19	Main engine switchboard
5	Main indicator light	20	Main engine starter switch
6	Bilge pump switch	21	Main engine choke control
7	Main engine ignition switch	22	Main engine oil pressure warning light
8	Trailer pump switch	23	Panel light switch
9	Tail or convoy light switch	24	Panel light cover
10	Charging set indicator light	25	Speedometer trip re-set
11	Magneto test switch	26	Oil pressure and fuel level gauge
12	Charging set ignition switch	27	Fuel level, oil pressure change-over switch
13	Charging set starter switch	28	Tachometer
14	Charging set switchboard		
15	Charging set engine throttle		

The instrument panel and ancillary controls in the driver's compartment of the FV4002 bridgelayer. (*Warehouse Collection*)

The FV4016 Centurion armoured ramp carrier (ARK) seen in travelling configuration. The ARK carries folding ramps at each end of the hull, with a trackway fitted across the top, and was designed to be driven into the centre of a ditch or trench, or up against a sea wall, before the ramps were folded out hydraulically at either end to form a Class 80 continuous bridge. The hull of the tank itself formed the centre section. This example is based on a Mk 3 or Mk 5 hull. (*Warehouse Collection*)

The Royal Netherlands Army operated a Centurion-based bridgelayer of its own design, mounting a forward-launched scissor bridge similar to that fitted to the US Army's M48 and M60 bridgelayers.

Churchill tanks had also been modified to act as armoured ramp carriers (ARKs) – effectively, a rapid assault bridge – during the Second World War and, although these also remained in service into the immediate post-war years, by the late 1950s they were showing their age. The FVRDE-designed replacement (FV4016) used the hull of a Centurion Mk 5 from which the turret and gun were removed, with a roof plate covering the turret ring; the commander was relocated inside the hull alongside the driver. Folding ramps were attached to either end of the hull, and a trackway was fitted across the top. In use, the tank was driven into the centre of a ditch or trench, or up against a sea wall, and the ramps were folded out hydraulically at either end to form a Class 80 continuous bridge, with the hull of the tank itself forming the centre section. In travelling configuration, the length of the ARK exceeded 34ft and, when deployed, the 81ft-long bridge gave a useful span of 75ft and a width of 14ft across the ramps.

There was some concern that the increased weight of the ARK would place greater loadings on the rear wheel stations and development trials for the ARK were carried out in 1957, with particular attention paid to the suspension loadings. Both Mk 5 (although in actual fact it was a Mk 3 hull that had been weighted to simulate a Mk 5) and Mk 7 hulls were trialled, with the objective of determining whether or not the performance of the vehicle was reduced by an unacceptable degree. The report published at the conclusion of the trials stated that the convoy performance of the Centurion ARK would be reduced by some 50–75 per cent when compared to the gun tank, and that tyre temperatures would become critical after running for 90 minutes, as opposed to 120–150 minutes for the latter.

The Centurion ARKs, which were constructed by ROF Leeds, remained in service until 1975, by which time this type of equipment was considered to be obsolete.

Another variant of the ARK was the so-called Centurion ARK mobile pier (CAMP), which was designed to provide a central pier support in the centre of a waterway that could accept two 'number 6' tank bridges. The ramps were omitted, leaving only the central trackway in place.

During the Second World War Churchill tanks had also been converted to the 'armoured vehicle, Royal Engineers' (AVRE) role, being used for the demolition of structures such as pillboxes and gun emplacements, and for carrying a fascine bundle that could be used to allow ditch crossing. Development of the Churchill AVRE continued into the post-war years, but by the end of the decade attention had switched to the Centurion as a basis for such a vehicle.

The first proposals for a Centurion-based AVRE date back to September 1950, when a prototype was produced at the Fighting Vehicles Proving Establishment (FVPE) using the hull of Centurion prototype number four (T352413, subsequently renumbered 02BA58), and the turret of a Centurion Mk 1 into which had been installed a 95mm howitzer. A full-width (13ft) hydraulically operated compact 'dozer blade, produced by the Newcastle company T.B. Pearson & Sons, was fitted to the nose, with a cradle to carry the fascine bundle above it. These changes effectively shifted the centre of gravity of the 53-ton vehicle forward of the centreline of the suspension by around 6in. In 1954/55 there were further trials with a Centurion Mk 3 (02BA12) that had been weighted to represent a Centurion AVRE carrying a 5-ton fascine bundle on the cradle, bringing the total weight up to 55.6 tons. The trials were intended to report on any negative impacts on the suspension and track system, and the vehicle was tested across 146 miles, with the speed limited to 12mph on the road and 8mph across country. It was concluded that carrying the fascine bundle 'causes no apparent damage to the suspension of the vehicle'. Trials were also carried out with a modified Mk 3-based AVRE that had been fitted with simplified gun controls.

The Centurion armoured vehicle Royal Engineers' (AVRE) first appeared in 1950, with prototypes subsequently constructed using the hulls of Mk 1, Mk 3 and Mk 7 gun tanks. When the vehicle went into production, it was based on the Mk 5 hull and was fitted with a 165mm L9A1 demolition gun, a full-width 'dozer blade, a hydraulic winch and a cradle for carrying a fascine bundle. (*Warehouse Collection*)

Despite all these trials, problems with development and a shortage of materials meant that the final prototype, by now using the hull of the Centurion Mk 7 gun tank, did not appear until August 1957. By this time the vehicle featured a front-mounted 'dozer blade, the rams for which were fitted into armoured boxes on the nose plate; a removable 10-ton jib, eventually mounted on a ball and socket arrangement; a cradle for carrying a fascine bundle, with the release effected via blow-out pins; and a 1.5-ton hydraulically operated winch. Hydraulic controls for the 'dozer and winch were provided in the driver's compartment. Definitive development trials, concentrating on the hydraulics and the ancillary equipment, were initiated in 1959 but there were further delays before production finally got under way in 1963. (See *colour plate 8, a and b.*) By this time a shortage of Mk 7s meant that the production vehicle was to be based on the hull of the Mk 5, into which was fitted a 165mm L9A1 gun firing a heavy demolition charge, but lacking the automatic stabiliser equipment found on the gun tanks. There was also an L3A3 or L3A4 0.30in Browning machine gun – the use of this gun on the AVRE was something of an anomaly since it remained in service for this application long past

A privately owned Centurion AVRE seen at the annual War & Peace Show. Although lacking finesse, the stubby-barrelled 165mm L9A1 gun was useful for demolishing structures such as pillboxes and gun emplacements. Note the additional armour on the hull and turret. (*Simon Thomson*)

its demise elsewhere in the British Army. Just forty vehicles were constructed, some of which were based on the hulls of Centurion Mk 12 artillery observation posts fitted with a Pearson's mine plough in place of the standard 'dozer blade, and retaining the 105mm gun of the original. A jettisonable 15-ton four-wheel trailer (FV2721A) was also developed for use with the Centurion AVRE to carry either a fascine bundle or trackway.

Although the AVRE was equipped with a hydraulic 'dozer blade, back in 1958 a requirement had been issued for fitting such a blade to a Centurion gun tank, but it was to be a further three years or so before the vehicle (FV4019) started to enter service. Designed to replace the Churchill- and Centaur-based 'dozers, the latter lacking a turret, the Centurion 'dozer was based on the Mk 5 gun tank to which had been fitted a hydraulically operated bulldozer blade manufactured by Pearson's and identical to that fitted to the Centurion AVRE. The blade was operated via a hydraulic pump manufactured by H.M. Hobson, and the equipment was supplied as a prefabricated 'dozer kit that could be fitted to either the Centurion or the Conqueror, with the hydraulic pump driven by the main engine. The main gun remained operative, although there was some reduction in the number of rounds that were carried, and both Mk 5 tanks, equipped with the 20-pounder (83.4mm) gun, and Mk 5/1 tanks with the 105mm L7 gun, were converted. With the 'dozer equipment in place, the weight was increased to 52.6 tons, and the tank became somewhat nose heavy and unstable when firing to the rear; the additional weight also precluded up-armouring.

There were initial problems with hydraulic connections, but it was a useful piece of kit, and the report of the development trials stated categorically that the 'dozer-equipped Centurion was 'the first armoured bulldozer to exceed by a considerable margin the earth-moving capacity of the Caterpillar D8' (bulldozer), shifting 300yd^3 of soil an hour compared to just 200 for the Caterpillar, and it was able to dig a 'hull-down' defensive position for a tank in light soil in just 7 minutes. It was also capable of removing concrete and steel road-block obstacles, hardwood trees up to around 36in diameter, and of breaching reinforced concrete walls.

Centurion 'dozers were also exported to Australia and Denmark.

Chapter Eight

Centurion Projects and Oddballs

Given the reliability and robustness of its hull and automotive components, it is hardly surprising that the Centurion was chosen as the basis for any number of projects and one-offs. Some of these never passed beyond the paper stage: FV4001, for example, was described as a mine clearer, and FV4019 was to be a flame-thrower variant but there is little evidence that anything was actually constructed for either project. However, in other instances, prototypes were actually constructed before the particular project was abandoned.

A deep-wading kit was developed for the Centurion that allowed immersion in up to 108in of water. Trials were carried out during 1957 using a Mk 7 (42BA34) with huge air stacks and an inflatable turret-ring seal. There were problems with the explosive bolts designed to allow the air stacks to be jettisoned, and the trials also showed that it was not possible for a Centurion coupled to the mono-wheeled trailer to disembark from the deck of a tank landing craft (LCT8) without compromising the fuel and electrical connections between the tank and the trailer unless the lengths of fuel line and cable were increased and the couplings waterproofed. Following a series of modifications, the wading kit was considered to be satisfactory ... but it seems never to have been used.

During 1954/55 trials were also conducted on a Centurion Mk 3 (00BA95) that was fitted with a duplex-drive (DD) system and a deep-water canvas screen that would allow the tank to be launched from a landing craft at some distance from the beach. Like that fitted to the DD Sherman tanks of the Second World War, the equipment was developed by the Hungarian inventor Nicholas Straussler working with FVRDE, and essentially consisted of a flotation screen, a twin-screw propeller drive system, a compressor and bilge pumps. The flotation screen was manufactured from rubberised Egyptian cotton, and was mounted on a sheet metal decking running around the perimeter of the tank at the level of the track guards; it was erected by a series of compressed-air struts and locked into position manually. The propellers, which were mounted at the rear, were driven by the sprockets, and were

A Centurion Mk 3 fitted with snorkel equipment. The large diameter tube provided air for the crew and the carburettors, while the smaller diameter tube at the rear was designed to vent petrol fumes from the engine compartment. (*Tank Museum*)

also raised and lowered by compressed air; the propeller mountings could be rotated in a horizontal plane to provide steering in the water. It was concluded that the Centurion could thus be made amphibious, and plans were made to produce DD Centurions in both Mk 5 and Mk 7 form (FV4008). However, despite considerable to-ing and fro-ing between FVRDE, the Amphibious Warfare Experimental Establishment (AWXE) and whichever of Straussler's companies was responsible, the various issues raised during the development trials were never satisfactorily resolved and the scheme was eventually abandoned.

There was a further attempt to make the Centurion swim in the mid-1960s, using a dozen rigid rather than flexible panels, but this too was abandoned, as was the pneumatic ring designed to encircle the hull and allow the tank to float.

A snorkel kit was developed for the Centurion in the 1960s, allowing it to wade across water obstacles fully submerged. The kit consisted of a PVC cover for the engine deck, covers for the driver's hatch, and a snorkel that was designed to fit over the commander's cupola; combustion air was drawn through the fighting compartment and exhaust gases were discharged through a ball-and-cage fitting on the exhaust pipe that prevented the ingress of water. Modifications were required when it was discovered that petrol vapour accumulated in the engine compartment, creating a fire risk.

In the early 1950s several proposals were made to use the Centurion chassis as a self-propelled (SP) gun mount. For example, in 1951 a 180mm gun was mounted on a modified Centurion Mk 3 hull, at first with an open top and with the gun having a limited traverse, a concentric recoil system and an auto-loader, and latterly with the gun mounted in a light, splinter-proof enclosed turret, with a conventional recoil system. Although the original 180mm gun was subsequently replaced by a 183mm weapon, the project, designated FV4005, never progressed beyond the basic feasibility stage and by August 1957 had been abandoned without reaching the production stage. FV3802 was a 1955 attempt at using a shortened version of the Centurion Mk 7 hull, equipped with the OQF 25-pounder (87.6mm) gun. The vehicle was not considered satisfactory and, under pressure from the Royal Artillery, evolved into the FV3805, which mounted a 5.5in gun. The project was eventually cancelled in 1960 in favour of the FV433 Abbott, which had a 105mm gun. There were also plans to use the Centurion chassis to mount a 7.2in howitzer (FV3806), a 120mm anti-tank gun (FV3807), a 20-pounder (84mm) medium gun (FV3808), and a 155mm gun (FV3809). None progressed beyond the stages of feasibility discussions and/or mock-ups, and all were quickly discounted.

Between about 1967 and 1971 British Aerospace tried to interest export customers in a Centurion Mk 5 that had been equipped with the British Aircraft Corporation (BAC) Swingfire wire-guided anti-tank missile system (ATGW).

A Centurion Mk 7 with metal trunking to allow wading; this equipment was fitted to the Centurions that waded ashore during the Suez affair. The muzzle cover was designed to be punctured by the first round fired from the main gun, while the canvas covers over the driver's hatch and the smoke grenade launchers were jettisoned by means of cordite charges. (*Warehouse Collection*)

Dating from 1951, the FV4005 'Stage One', described as a 'self-propelled heavy anti-tank gun', mounted a 183mm auto-loading gun in an open-topped Centurion Mk 3 hull; the gun traverse was limited. This was superseded by the FV4005 Stage Two. (*Warehouse Collection*)

(*Above*) For the FV4005 'Stage Two', the Centurion Mk 3 hull was retained but the 183mm weapon was installed in a lightly armoured splinter-proof housing. In this configuration the gun was loaded by hand, but the project never progressed beyond the basic feasibility stage and had been abandoned by August 1957 without reaching the production stage. (*Warehouse Collection*)

(*Opposite, top*) The FV3805 was a self-propelled 5.5in howitzer mounted on what was basically a heavily modified Centurion Mk 7 hull. The gun was mounted on a turntable that permitted traversing to 30 degrees either side of the vehicle centreline, combined with a maximum elevation of 70 degrees. Two prototypes were constructed in 1956, but the project was abandoned in 1960. (*Warehouse Collection*)

(*Opposite, below*) Three Centurion Mk 3s were converted to moving target tanks by REME 27 Command Workshops for training anti-tank crews in the use of the MILAN anti-tank wire-guided missile, and the 120mm WOMBAT and 84mm Carl Gustav recoilless rifles. Although live ammunition was never involved during such training, the additional armour brought the weight of the 'target' up to 60 tons! (*Warehouse Collection*)

Although it never progressed beyond the mock-up stage, the company showed how twin missile launchers could be mounted on each side of the standard turret, with two spare missiles carried on the turret rear. The main gun was unaffected. And for those customers considering the Centurion as the basis of an anti-aircraft weapons platform, the vehicle was also proposed as a mount for the Marconi Marksman anti-aircraft turret.

Although it was never offered as a standard Centurion variant, during the 1960s at the tank gunnery range at Kirkcudbright in Scotland an obsolete Centurion Mk 2 was stripped of its turret and fitted with a Coles S7-10 fixed-jib crane to assist with handling gun mountings, targets and similar equipment on the ranges. When this device was damaged beyond repair by falling into a pit, the crane was replaced by a more modern Jones KL 11.7 with a capacity of 12 tons.

In Italy the Centurion chassis, as well as the American Patton M47, M48 and M60 tanks, was used by the Astra Company of Piacenza to develop a bridgelayer with a hydraulically launched Class 60 steel and aluminium scissors bridge. Small numbers were delivered to the Israeli Defence Force (IDF) and to the Italian Army, the latter almost certainly using the American chassis.

Finally, in Britain the Centurion was also used in various ways to assist in the development of the Chieftain main battle tank. For example, Centurion 'Action X' (84BA86) consisted of a Mk 7 hull mounting a 20-pounder gun in a new, low-profile cast turret that lacked a gun mantlet. Although the turret resembled that finally fitted to the Chieftain, it differed slightly from the final production version. Centurion automotive components were subsequently used to construct three examples of the FV4202 – sometimes described as the 40-ton Centurion – that was intended as a research vehicle to assist in the development of the supine driving position, the suspension and the turret of the Chieftain.

Three Centurions were converted to target tanks by REME 27 Command Workshops for training anti-tank crews in the use of the MILAN anti-tank wire-guided missile, and the 120mm WOMBAT and 84mm Carl Gustav recoilless rifles. Designed to be crewed by two men – a commander and a driver – the vehicles weighed some 60 tons and modifications included up-armouring of the hull, and fitting the tank with larger and more substantial side skirts; it is worth pointing out that live ammunition was never involved during such training! One of these curious contraptions remains at the Tank Museum.

Appendix

Centurion Reference Data

TECHNICAL SPECIFICATION

Typical nomenclature: 'tank, cruiser, Centurion; A41', then 'medium gun tank, Centurion; FV4000 series'.

Manufacturer: prototypes, AEC Ltd, Southall; Royal Ordnance Factory (ROF) Leeds, ROF Woolwich.

Production vehicles, ROF Dalmuir, Leeds, Nottingham and Woolwich; Vickers-Armstrongs Ltd (later, Vickers Defence Systems Ltd), Elswick, Newcastle-upon-Tyne; and Leyland Motors Ltd, Preston and Leyland, Lancashire.

Engines, Rover Company Ltd, Solihull, Warwickshire.

Total number produced: 4,423 (all variants).

Production: 1945–62.

Main engine: Rolls-Royce (Rover) Meteor Mk 4, 4A, 4B, 4B/1, 4B/1/H, or 4C; normally aspirated petrol engine; 12 cylinders in 60-degree V configuration; bore and stroke, 5.4in × 6in; 27,005cc (1,649in^3); compression ratio, 6:1 or 7:1; overhead valves, with four valves per cylinder; gross power output, 550–650bhp at 2,250–2,400rpm; maximum torque, 1,550lbf/ft at 1,600rpm.

Auxiliary engines: all variants, driving main generator; Morris USHNM Mk 2 or Mk 2/1; normally aspirated petrol engine; four cylinders; bore and stroke, 2.2in × 3.5in; 918cc (56in^3); side valves; gross power output, 20bhp at 2,500rpm; maximum torque, 33lbf/ft at 1,400rpm.

FV4002 bridgelayer, driving bridge-launching hydraulics; Rolls-Royce B40 No. 1 Mk 5P; normally aspirated petrol engine; four cylinders; bore and stroke, 3.5in × 4.5in; 2,838cc (173in^3); overhead inlet valves, side exhaust; gross power output, 62bhp at 2,800rpm; maximum torque, 209lbf/ft at 1,600rpm.

FV4006 armoured recovery vehicle (ARV), driving main winch; Rolls-Royce B80 No. 1 Mk 2P or 5P; normally aspirated petrol engine; eight cylinders in-line; bore and stroke, 3.5in × 4.5in; 5,675cc (346in^3); overhead inlet valves, side exhaust; gross power output, 165bhp at 3,750rpm; maximum torque, 280lbf/ft at 1,600rpm.

Transmission and steering system: Merritt-Brown Z51R constant-mesh controlled differential; 5F2R, manual shift; driving the rear sprockets through 16in triple dry-plate clutch.

Gearbox ratios: first gear, 11.643:1; second gear, 4.593:1; third gear, 2.855:1; fourth gear, 1.807:1; fifth gear, 1.343:1; reverse gear, 22.894:1; high-speed reverse gear, 3.859:1.

Final drive: double-reduction spur gear train; ratio, 7.47:1 (Mk 1 only, 6.94:1).

Turning radii: first gear, 15ft; second gear, 38ft; third gear, 61ft; fourth gear, 96ft; fifth gear, 130ft; reverse gear, 7.6ft; high-speed reverse gear, 45ft.

Suspension: AEC-Rackham (modified Horstman) system, consisting of 13.5in radial arms suspended on triple concentric horizontal coil springs, arranged as three bogies on either side; front and rear suspension units incorporating four hydraulic telescopic shock absorbers.

Brakes: Girling-Bendix mechanically operated, with 20in diameter drums attached to final drive shafts.

Road wheels: 31.6in diameter, twelve pairs each side.

Tracks: cast manganese steel, four track-return rollers; some tanks retro-fitted with track shoes having rubber pads.

Construction: welded steel hull, one-piece cast turret with welded roof; turret of Mk 1 fabricated from welded steel plate.

CAPACITIES

Fuel capacity: (Mk 3 and Mk 5) 121 gal; (Mk 7 and Mk 8) 230 gal.

Oil capacity: 14 gal.

Cooling system capacity: 33 gal.

Electrical system: 24V, wired on a negative earth system, using 4 x 6V 115Ah or 120Ah batteries.

PERFORMANCE

Maximum speed: on roads, 23.7–21.5mph; cross-country, 15mph.

Fuel consumption: 0.075 gal/bhp/hour, giving 0.83mpg on roads, 0.38mpg cross-country.

Maximum range: (Mk 3 and Mk 5) on roads, 62.5 miles; cross-country, 32.5 miles; (Mk 7 and Mk 8) on roads, 115 miles; cross-country, 60 miles.

Maximum gradient: 30–35 degrees at full tractive effort.

Vertical obstacle: 36in.

Trench crossing: 11ft.

Fording depth: unprepared, 42in; prepared, 110in (top of turret).

Maximum gun depression: 10 degrees (except over rear deck).

Maximum gun elevation: (Mk 3 and Mk 5) 18 degrees; (Mks 7 to Mk 9) 20 degrees.

DIMENSIONS AND WEIGHT

Mk 3 and Mk 5
Dimensions: length, 298in (hull only), 387in (gun forward), 339in (gun to the rear); width, 133in (with side skirts), 129in (without side skirts); height, to top of turret, 115in.
Ground clearance: 20in.
Track width: 24in wide (20in on prototypes), 5.5in pitch.
Track centres: 104in.
Length of track on ground: approximately 15ft.
Combat weight: 111,966lb.
Power to weight ratio: approximately 12.7bhp/ton.
Bridge classification: 60.

Thickness of armour: hull glacis plate, 76mm; hull nose, 76mm; hull sides, front, 51mm; hull sides, upper rear areas, 38mm; hull sides, lower rear areas, 20mm; floor, 17mm; turret front, 152mm.

Mk 7 and Mk 8
Dimensions: length, 299in (hull only), 389in (gun forward), 339in (gun to the rear); width, 133in (with side skirts), 129in (without side skirts); height, to top of turret, 115–117in.
Ground clearance: 20in.
Track width: 24in wide, 5.5in pitch.
Track centres: 104in.
Length of track on ground: approximately 15ft.
Combat weight: 114,240lb.
Power to weight ratio: approximately 12.5bhp/ton.
Bridge classification: 60.

Thickness of armour: hull glacis plate, 118mm; hull nose, 76mm; hull sides, front, 51mm; hull sides, upper rear areas, 38mm; hull sides, lower rear areas, 20mm; floor, 17mm; turret front, 152mm.

Table 1. Typical Ammunition Stowage

Mark	Main gun: 17-pounder	20-pounder	105mm	Secondary weapons: 7.92mm Besa	0.30in Browning	0.50in Browning
Mk 1	74 rounds	—	—	3,375 rounds	—	—
Mk 2	81 rounds	—	—	3,600 rounds	—	—
Mk 3	—	65 rounds	—	3,600 rounds	—	—
Mk 5	—	64 rounds	—	—	4,250 rounds	—
Mk 6	—	—	70 rounds	—	3,000 rounds	—
Mk 7	—	63 rounds*	—	—	3,000 rounds	—
Mk 8	—	63 rounds*	—	—	3,000 rounds	—
Mk 9	—	—	70 rounds	—	3,000 rounds	600 rounds
Mk 10	—	—	70 rounds	—	3,000 rounds	600 rounds
Mk 11	—	—	70 rounds	—	3,000 rounds	600 rounds
Mk 12	—	—	70 rounds	—	3,000 rounds	600 rounds
Mk 13	—	—	70 rounds	—	4,750 rounds	600 rounds

* 61 rounds when heater fitted.

Table 2. Major Centurion Variants

Until 1948 it was the practice of the War Office to use Roman numerals for mark numbers which means, for example, that the Mk 3 would have been referred to as the Mk III and so on.

Mark and identifier		Description	Main gun	Secondary weapon(s)
Prototypes				
—	A41	Cruiser tank; prototypes one to ten (P1–P10)	17-pounder (76.2mm)	coaxial 20mm Polsten cannon; 7.92mm Besa machine gun
—	A41	Cruiser tank; prototypes eleven to fifteen (P11–P15)	17-pounder (76.2mm)	coaxial 20mm Polsten cannon
—	A41	Cruiser tank; prototypes sixteen to twenty (P16–P20)	77mm	coaxial 7.92mm Besa machine gun
Production variants				
Mk I	A41*	Gun tank; rolled-steel turret	17-pounder (76.2mm)	coaxial 7.92mm Besa machine gun (some with 20mm Polsten cannon)
Mk I ARV	—	Armoured recovery vehicle; based on Mk I and Mk 2 hulls	n/a	0.30in Browning machine gun
Mk 2	A41A	Gun tank; cast turret	17-pounder (76.2mm)	coaxial 7.92mm Besa machine gun
Mk 2 ARV	FV4006	Armoured recovery vehicle; based on Mk I, 2, 3 and 5 hulls	n/a	0.30in Browning machine gun
Mk 3	—	Gun tank	20-pounder (83.4mm)	coaxial 7.92mm Besa machine gun
Mk 3 AVLB	FV4002	Bridgelayer; based on Mk 3 hull	n/a	n/a
Mk 5	FV4011	Gun tank	20-pounder (83.4mm)	2x 0.30in Browning machine guns, one coaxial, one anti-aircraft
Mk 5/1	FV4011	Gun tank; up-armoured Mk 5	20-pounder (83.4mm)	2x 0.30in Browning machine guns, one coaxial, one anti-aircraft
Mk 5/2	—	Gun tank; as Mk 5, with additional fuel capacity and 105mm L7 gun	105mm L7	2x 0.30in Browning machine guns, one coaxial, one anti-aircraft
Mk 5 'dozer	FV4019	Gun tank, as Mk 5, with hydraulic 'dozer blade	20-pounder (83.4mm)	2x 0.30in Browning machine guns, one coaxial, one anti-aircraft
Mk 5/2 'dozer	FV4019	Gun tank, as Mk 5/1, with hydraulic 'dozer blade and 105mm L7 gun	105mm L7	2x 0.30in Browning machine guns, one coaxial, one anti-aircraft
Mk 5/3 'dozer	FV4019	Gun tank, as Mk 5/2, with ranging machine gun, and hydraulic 'dozer blade	105mm L7	2x coaxial Browning machine guns, one 0.30in, one 0.50in
Mk 5 ARK	FV4016	Armoured ramp carrier; based on Mk 5 hull	n/a	n/a
Mk 5 AVLB	FV4002	Bridgelayer; based on Mk 5 hull	n/a	n/a

Variant	FV number	Description	Gun	Armament
Mk 5 AVRE	FV4003	Armoured vehicle, Royal Engineers; based on Mk 5 hull	165mm BL	n/a
Mk 5 BARV	FV4018	Beach armoured recovery vehicle; based on Mk 1, Mk 2 and Mk 5 hulls	n/a	n/a
Mk 6	–	Gun tank; up-armoured Mk 5, with long-range fuel tanks	105mm L7	2x 0.30in Browning machine guns, one coaxial, one anti-aircraft
Mk 6/1	–	Gun tank; as Mk 6, with infrared night-vision equipment, and turret stowage basket	105mm L7	2x 0.30in Browning machine guns, one coaxial, one anti-aircraft
Mk 6/2	–	Gun tank; as Mk 6, with ranging machine gun	105mm L7	2x coaxial Browning machine guns, one 0.30in, one 0.50in
Mk 6/2 LR	–	Gun tank; as Mk 6, with ranging machine gun and long-range fuel tanks	105mm L7	2x coaxial Browning machine guns, one 0.30in, one 0.50in
Mk 7	FV4007	Gun tank; redesigned by Leyland Motors with enlarged hull and fume extractor	20-pounder (84mm)	2x 0.30in Browning machine guns, one coaxial, one anti-aircraft
Mk 7/1	FV4012	Gun tank; up-armoured Mk 7	20-pounder (84mm)	2x 0.30in Browning machine guns, one coaxial, one anti-aircraft
Mk 7/2	–	Gun tank; as Mk 7, with 105mm L7 gun	105mm L7	2x 0.30in Browning machine guns, one coaxial, one anti-aircraft
Mk 8	FV4014	Gun tank; as Mk 7/2; new commander's cupola and resilient gun mantlet	20-pounder (84mm)	2x 0.30in Browning machine guns, one coaxial, one anti-aircraft
Mk 8/1	FV4007	Gun tank; up-armoured Mk 8	20-pounder (84mm)	2x 0.30in Browning machine guns, one coaxial, one anti-aircraft
Mk 8/2	–	Gun tank; as Mk 8, with 105mm L7 gun	105mm L7	2x 0.30in Browning machine guns, one coaxial, one anti-aircraft
Mk 9	FV4015	Gun tank; up-armoured Mk 7, with 105mm L7 gun	105mm L7	coaxial .30in Browning machine gun
Mk 9/1	FV4007	Gun tank; as Mk 9, with infrared night-vision equipment, and turret stowage basket	105mm L7	coaxial .30in Browning machine gun
Mk 9/2	FV4007	Gun tank; as Mk 9, with ranging machine gun	105mm L7	2x coaxial Browning machine guns, one 0.30in, one 0.50in
Mk 10	FV4017	Gun tank; up-armoured Mk 8, with 105mm L7 gun	105mm L7	coaxial .30in Browning machine gun
Mk 10/1	–	Gun tank; as Mk 10, with infrared night-vision equipment and turret stowage basket	105mm L7	coaxial .30in Browning machine gun
Mk 10/2	–	Gun tank; as Mk 10, with ranging machine gun	105mm L7	2x coaxial Browning machine guns, one 0.30in, one 0.50in
Mk 11	–	Gun tank; as Mk 6, with ranging machine gun, infrared night-vision equipment and turret stowage basket	105mm L7	2x coaxial Browning machine guns, one 0.30in, one 0.50in
Mk 12	FV4007	Gun tank; as Mk 9, with ranging machine gun, infrared night-vision equipment and turret stowage basket	105mm L7	2x coaxial Browning machine guns, one 0.30in, one 0.50in

Mark	FV number	Description	Main armament	Secondary armament
Mk 13	—	Gun tank, as Mk 10, with ranging machine gun and infrared night-vision equipment	105mm L7	2x coaxial Browning machine guns, one 0.30in, one 0.50in

Experimental and development variants, many of which never progressed beyond mock-up stage

Mark	FV number	Description	Main armament	Secondary armament
P8	A41	Canal defence light mounted on prototype number eight	17-pounder (76.2mm)	coaxial 20mm Polsten cannon; 7.92mm Besa machine gun
Mk 1	FV4001	Mine clearer	n/a	n/a
Mk 2	FV4006	Spare gun barrel carrier	—	—
Mk 3	FV4004	Conway interim heavy gun tank based on Mk 3 hull	120mm L1	n/a
Mk 3	—	Flame-thrower	20-pounder (83.4mm)	flame projector; coaxial 7.92mm Besa machine gun
Mk 4	—	Close support (CS) tank; development only	95mm howitzer	coaxial 7.92mm Besa machine gun
Mk 5	—	Mount for Swingfire anti-tank missile	20-pounder (83.4mm)	2x 0.30in Browning machine guns, one coaxial, one anti-aircraft
Mk 6	FV4005/1	Self-propelled heavy anti-tank gun, limited traverse	183mm	n/a
Mk 6	FV4005/2	Self-propelled heavy anti-tank gun, turreted	183mm	n/a
Mk 5, Mk 7	FV4008	Gun tank with duplex-drive (DD)	20-pounder (83.4mm)	2x 0.30in Browning machine guns, one coaxial, one anti-aircraft
—	FV3801	Gun tractor	n/a	n/a
—	FV3802	25-pounder self-propelled gun, based on Mk 7 hull	25-pounder (87.6mm)	n/a
—	FV3803	Command post vehicle	n/a	n/a
—	FV3804	Section ammunition vehicle	n/a	n/a
—	FV3805	5.5in self-propelled howitzer; based on Mk 7 hull	5.5in	n/a
—	FV3806	7.2in self-propelled gun	7.2in	n/a
—	FV3807	120mm self-propelled anti-tank gun	120mm	n/a
—	FV3808	Self-propelled medium gun	n/a	n/a
—	FV3809	155mm self-propelled gun	155mm	n/a
—	FV4009	Self-propelled heavy anti-tank gun	7.2in	n/a
—	FV4010	Mount for Malkara anti-tank missile	n/a	n/a
AVRE Mk 3	FV4013	Armoured vehicle, Royal Engineers	n/a	n/a

Table 3. Number of Centurions Constructed
No production examples of the Mk 4 were built, and Centurions Mks 6, 9, 11, 12 and 13 were modified from existing vehicles of other 'marks'.

Date	Mk 1	Mk 2	Mk 3	Mk 5	Mk 7	Mk 8	Mk 9	Mk 10
1945/46	–	1	–	–	–	–	–	–
1946/47	48	57	–	–	–	–	–	–
1947/48	52	192	30	–	–	–	–	–
1948/49	–	–	139	–	–	–	–	–
1949/50	–	–	193	–	–	–	–	–
1950/51	–	–	229	–	–	–	–	–
1951/52	–	–	500	–	–	–	–	–
1952/53	–	–	573	–	–	–	–	–
1953/54	–	–	565	–	1	–	–	–
1954/55	–	–	359	–	154	–	–	–
1955/56	–	–	245	36	129	11	–	–
1956/57	–	–	–	176	168	51	–	–
1957/58	–	–	–	9	131	16	–	–
1958/59	–	–	–	–	78	16	–	–
1959/60	–	–	–	–	94	14	1	29
1960/61	–	–	–	–	–	–	–	110
1961/62	–	–	–	–	–	–	–	16
Totals	100	250	2,833	221	755	108	1	155

Grand total constructed: 4,423

Table 4. Centurion Contract Details
Table includes all available data but is not necessarily comprehensive.

Date	Contract number	Mark	Contractor	Quantity	Registration numbers
1945	n/a	Mk 1, Mk 2	ROF	110	00ZR03–01ZR12
1948	n/a	Mk 3	ROF	752	01ZR13–08ZR64
1949	6/FV/1132	Mk 3	ROF	40	00BA01–00BA40
1949	6/FV/1138	Mk 3	Vickers-Armstrongs	51	00BA41–01BA91
1949	6/FV/2006	Mk 3	Vickers-Armstrongs	60	04BA36–04BA95
1950	6/FV/3214	Mk 3	Vickers-Armstrongs	82	04BA96–05BA77
1950	6/FV/3215	Mk 3	n/a	189	05BA78–07BA66
1950	6/FV/3891	Mk 3, Mk 5	ROF	650	09BA28–13BA04, 13BA05–15BA77
1950	6/FV/3892	Mk 3	Vickers-Armstrongs	350	15BA78–19BA27
1951	6/FV/5493	Mk 3	Leyland Motors	400	42BA25–46BA24
1951	6/FV/5544	Mk 3	ROF	530	46BA25–50BA14, 71BA47–72BA09, 88BA20–88BA82*, 01BB93–02BB07
1952	6/FV/8340	Mk 3	ROF	327	60BA11–62BA10, 70BA20–71BA46
1953	6/FV/10209	Mk 7	Vickers-Armstrongs	33	84BA49–84BA81
1953	6/FV/10213	Mk 7	ROF	120	84BA82–85BA46, 91BA56–92BA10
1954	6/FV/10102	ARV Mk 2	ROF	17	99BA35–99BA51
1954	6/FV/10209	Mk 7, Mk 8	Vickers-Armstrongs	23	91BA33–91BA55, 09BB88
1954	6/FV/10968	ARV Mk 2	ROF	8	99BA52–99BA59
1954	6/FV/11133–4	ARV, ARV Mk 2	Vickers-Armstrongs	28	88BA83–89BA07, 01BB53–01BB55
1954	6/FV/11135	ARV Mk 2	ROF	35	89BA08–89BA32, 96BA41–96BA50
1954	6/FV/11136	Mk 5	n/a	134	89BA99–91BA32
1954	6/FV/13342	ARV Mk 2	n/a	16	01BB57–01BB72
1954	6/FV/11147	ARV	ROF	25	89BA75–89BA98, 06BB22
1954	6/FV/11148	ARV	Vickers-Armstrongs	40	89BA33–89BA74
1954	6/FV/11965*	Mk 3	Vickers-Armstrongs	14	93BA59–93BA82
1954	6/FV/11986	Mk 3	ROF	98	93BA83–94BA70, 94BA71–94BA80*
1954	6/FV/12719	Mk 2 ARV	Vickers-Armstrongs	9	00BB42–00BB50
1954	6/FV/12738	Mk 5	ROF	37	99BA60–99BA96
1954	6/FV/12739	Mk 5	ROF	14	99BA97–99BA99, 00BB01–00BB11
1954	6/FV/12954	ARV Mk 2	n/a	9	01BB44–01BB52

Year	Contract	Mark	Manufacturer	Quantity	Serial range
1955	6/FV/16628	Mk 8	ROF	32	09BB20–09BB51
1956	6/FV/15752	Mk 7	ROF	2	09BB86–09BB87
1959	6/FV19665	Mk 10	ROF	19	04CC78–05CC06
1959	KL/A/049	Mk 10	Vickers-Armstrongs	1	00DA01
1960	KL/A/0136	Mk 10	Vickers-Armstrongs	6	02DA88–02DA93
1960	KL/A/0147	Mk 10	ROF	10	02DA94–03DA03

* Contract (or part quantities of contract) cancelled.

Table 5. Centurion Users

Nation	Date	Number	Notes
Australia	1950	144	Including some surplus vehicles from Britain and New Zealand
Canada	1952	374	
Denmark	1952	216	Mk 3s supplied under US government MDAP arrangements*; total also includes some surplus vehicles from Britain
Egypt	1956	30	
Great Britain	1945	±2,000	
India	1955	245	
Iraq	1965	120	
Israel	1959	±1,000**	Including surplus vehicles from Britain and Canada; some vehicles subsequently upgraded and renamed Sho't and Sho't Kal
Jordan	1954	293	Subsequently upgraded and designated Tariq
Kuwait	1961	16	
Libya	1957	10	
Netherlands	1953	1,065	Mk 3s supplied under US government MDAP arrangements*
New Zealand	1952	12	
Singapore	n/a	12–63	Exact number not known; bought from India and Israel; designated Tempest
Somalia	n/a	30	
South Africa	1953	541	Including surplus vehicles from India and Jordan; some vehicles subsequently upgraded and renamed Skokiaan, Semel and Olifant
Sweden	1952	350	Designated Strv 80 (Stridsvagn), Strv 101, Strv 102 and Strv 104
Switzerland	1955	430	Including some surplus vehicles from South Africa; designated Panzer 55 (Pz 55, Pz 55/60) and Panzer 57 (Pz 57, Pz 57/60)

* Centurions operated by Denmark and the Netherlands were purchased from the British government by the Economic Co-operation Agency of the Foreign Service of the USA, and were supplied to the user nations under the Mutual Defense Assistance Program (MDAP) initiated in 1949. These vehicles remained the property of the US government.
** The Israeli Defence Force (IDF) originally purchased around 250 new Centurions.